I0482840

Proceedings of the Second International Fishing Industry Safety and Health Conference

EDITED BY

Nicolle A. Mode, MS
Priscilla Wopat, MA
George A. Conway, MD, MPH

September 22-24, 2003
Sitka, Alaska, U.S.A.

Convened by
U.S. Department of Health and Human Services
Public Health Service
Centers for Disease Control and Prevention
National Institute for Occupational Safety and
Health

AND

Alaska Marine Safety Education Association

April 2006

DISCLAIMER

Sponsorship of the IFISH II Conference and these proceedings by the National Institute for Occupational Safety and Health (NIOSH) does not constitute endorsement of the views expressed or recommendations for use of any commercial product, commodity, or service mentioned. The opinions and conclusions expressed in the papers are those of the authors and not necessarily those of NIOSH.

Recommendations are not to be considered as final statements of NIOSH policy or of any agency or individual involved. They are intended to be used in advancing the knowledge needed for improving worker safety and health.

This document is in the public domain and may be freely copied or reprinted.

Copies of this and other NIOSH documents are available from
Publications Dissemination, EID
National Institute for Occupational Safety and Health
4676 Columbia Parkway
Cincinnati, OH 45226-1998

Fax number: (513) 533-8573
Telephone number: 1-800-35-NIOSH (1-800-356-4674)
e-mail: pubstaft@cdc.gov

For further information about occupational safety and health topics, call 1-800-35-NIOSH (1-800-356-4674), or visit the NIOSH Web site at
www.cdc.gov/niosh

DHHS (NIOSH) PUBLICATION No. 2006-114

Table of Contents

Table of Contents (Continued)

Table of Contents (Continued)

Foreword

Commercial fishermen continue to risk their lives and livelihood as they labor to bring food to tables around the world. Few occupations are as dangerous as that of a commercial fisherman's, and we at the National Institute for Occupational Safety and Health place the safety of these workers as a high priority. We call upon the readers of this proceedings volume to join our efforts to support safety training for commercial fishermen and the acquisition and use of safety equipment, including personal flotation devices, survival suits, and radio equipment, for all commercial fishing vessels. While we may not be able to control the harsh environment in which commercial fishing takes place, we certainly can promote safer vessels and survival training for workers in the commercial fishing industry.

John Howard, MD, MPH, JD, LLM
Director
National Institute for Occupational Safety and Health
Washington, DC

Acknowledgments

We appreciate the help provided to us by Kristie Sherrodd, manager, Alaska Marine Safety Education Association (AMSEA), who helped organize the conference and helped editors track down material for these proceedings; Jerry Dzugan, executive director, AMSEA, who helped organize the conference and facilitated communications with many of the presenters at the conference; Diana Hudson, Angela Hunt, and Michael Jones, who gathered and organized these articles and completed some of the early editorial work; Jennifer Lincoln, Alaska Field Station, who assisted with final edits and layout; Linda Bradford, Alaska Field Station, who assisted with layout and editing; and Tim Pizatella, NIOSH Division of Safety Research, who assisted with editing an earlier version of the proceedings. Lance Kissler of the Spokane Research Laboratory was invaluable in providing assistance with the InDesign document program. We also thank Regina Pana-Cryan, NIOSH senior scientist, for her review of the document. We gratefully acknowledge the help provided by AMSEA staff during the conference, including Kris Finkenbinder, Mary Chambers, Steven Campbell, and Julie Butler Doggett. We also appreciate the individual presenters at the Second Conference on International Fishing Industry Safety and Health in Sitka, Alaska, who shared their expertise in commercial fishing safety issues with attendees. Most of the presentations featured at IFISH II are available within the pages of this volume, and we believe the material presented here will help promote commercial fishing safety around the world.

Scientific Committee

Ann Backus, MS, Harvard School of Public Health, USA
David Claribut, Workers' Compensation Board of British Columbia, Canada
George Conway, MD, MPH, CDC/NIOSH, Alaska Field Station, USA (IFISH II co-chair)
Jerry Dzugan, MS, Director, Alaska Marine Safety Education Association, USA (IFISH II co-chair)
Richard Hiscock, USA
Diana Hudson, CDC/NIOSH, Alaska Field Station, USA (IFISH II co-chair)
Brad Husberg, CDC/NIOSH, Alaska Field Station, USA
Sue Jorgensen, Fishing Vessel Safety Coordinator, US Coast Guard
Michael Murray, PhD, Memorial University of Newfoundland, Canada
Gudrun Pedursdottir, PhD, Fisheries Research Institute, University of Iceland
Hannu Rintamaki, PhD, Regional Institute of Occupational Health, Finland
Thomas Smith, PhD, MPH, MS, CIH, Harvard School of Public Health, USA
Jeremy Turner, BSc, Senior Fishery Industry Officer, FAO, Italy
Brandt Wagner, Maritime Specialist, International Labour Office, Switzerland

International Committee

Menakhem Ben Yami, Advisor, Fisheries Development and Management, Israel
George Conway, MD, MPH, CDC/NIOSH, Alaska Field Station, USA (Committee Chair)
Robert Ekman, PhD, Karolinska Institute, Sweden
Leslie Hughes, Executive Director, North Pacific Vessel Owners Association, USA
Olaf Jensen, MD, MPH, University of Southern Denmark, Denmark
Sammy Park, Senior Investigator, Korean Maritime Safety Tribunal, South Korea
Eduardo Rossel, Coordinator, National Fishery Program, Chile
Kristie Sherrodd, ML, Alaska Marine Safety Education Association, USA
Al Steinman, MD, MPH, Marine Safety Advocate, USA
Jeremy Turner, BSc, Senior Fishery Industry Officer, FAO, Italy

Preface

Fatal traumatic injuries in commercial fishing have resulted in this industry being one of the most hazardous in Alaska, the United States, and many other nations. The International Labour Organization (ILO) and Food and Agriculture Organization (FAO) estimate that 7% of all worker fatalities worldwide occur in the fishing industry, even though this industry accounts for less than 1% of the worldwide workforce. The fatality rate for U.S. commercial fishermen was 168 per 100,000 workers per year from 1994 through 1998, 35 times the overall US occupational fatality rate (4.8 per 100,000 workers per year) (CFOI). Around the world, for example, in Australia, Denmark, Finland, Korea, and Sweden, occupational fishing fatality rates range from 16 to as much as 79 times higher than these countries' overall occupational fatality rate. The ILO has estimated that the fishing industry experiences 24,000 deaths and as many as 24 million nonfatal injuries each year worldwide.

To bring together fishermen, fishing safety proponents and professionals, government officials, equipment manufacturers, and other parties interested in fishing safety and health, the Alaska Field Station, National Institute for Occupational Safety and Health, organized the Fishing Industry Safety and Health (FISH) conferences. The first two (Anchorage, Alaska, in 1992, and Seattle, Washington, in 1997) were national in scope. As these were well-attended and included participants wanting to learn from other countries where fishing was of economic significance, we decided to broaden the scope of the next conference. Thus, the first International Fishing Industry Safety and Health Conference (IFISH) was held in Woods Hole, Massachusetts, in October of 2000, in collaboration with the Harvard School of Public Health. That meeting was well attended and included representatives from many nations.

In late September of 2003, working with the Alaska Marine Safety Education Association, we held IFISHII in Sitka, Alaska, which drew 135 registrants from 18 nations. Forty speakers addressed topics ranging from deck safety needs for crabbers working in northern waters to policy changes affecting Pacific Island States. Among the presenters were seven speakers sponsored by FAO who provided overviews of commercial fishing safety programs in developing countries, including Tonga, Sri Lanka,

Pakistan, India, Senegal, and Chile. IFISH II's focus on safer working environments for commercial fishermen is part of a growing international emphasis on the need for collaboration among governments, nongovernmental entities, vessel owners and operators, and fishermen themselves to develop effective safety programs. Although fishermen from Sri Lanka sometimes face different types of problems than do fishermen from Sweden or the United States, all of them are operating offshore, usually at some distance from emergency help.

This proceedings volume includes manuscripts submitted for 28 of the 40 presentations given at the conference. The range of subjects is impressive, from risk factor analyses to intervention approaches, some rooted in practicalities and success, some more theoretical. The presentations and resulting papers represent tremendous geographic diversity as well, with papers presented and submitted by fishermen from the South Pacific all the way to the Arctic Circle. Gathering people from fishing countries spread around the globe at an event like IFISH II helps us all to identify programs, equipment, and policies that are effective in promoting fishing safety.

George A. Conway, MD, MPH
National Institute for Occupational Safety and Health
Anchorage, Alaska
January 2004

Executive Summary

This volume contains material that was presented at the Second International Fishing Industry Safety and Health Conference in Sitka, Alaska, on September 22 through September 24, 2003. IFISH II was sponsored by NIOSHs Alaska Field Station and convened with the help of the Alaska Marine Safety Education Association of Sitka. More than 125 participants attended sessions on various aspects of commercial fishing safety. Forty speakers addressed topics ranging from deck safety needs for crabbers working in northern waters to policy changes for Pacific Island States. Among the presenters were seven speakers sponsored by the Food and Agriculture Organization who provided overviews of commercial fishing safety programs in developing countries, including Tonga, Sri Lanka, Pakistan, India, Senegal, and Chile.

An emphasis on practical solutions for fishing safety emerged in presentations given throughout the conference. Speakers discussed slip-resistant footwear, orientation and safety programs for new seafood processor workers from rural Alaska, and programs for increasing physical fitness for Faroese Islander fishermen. Other speakers focused on findings from recent fishing vessel disasters and discussed ways that fishing vessel owners and crew could prevent such events. All had the opportunity to present their findings and answer questions from the audience.

IFISH IIs focus on safer working environments for commercial fishermen is part of a growing international emphasis on the need for collaboration among governments, nongovernmental entities, vessel owners and operators, and fishermen themselves to develop effective safety programs. Our hope is that these proceedings contribute to these efforts by underscoring the hazards that commercial fishermen face each day at work and by illustrating the many ways in which collaborative partnerships can help promote safer fishing throughout the world.

Nicolle A. Mode, MS
Priscilla Wopat, MA
George A. Conway, MD, MPH

OPENING SESSION:
AN OVERVIEW OF MULTINATIONAL FISHING
SAFETY EFFORTS

Chilean fishermen (Photograph courtesy of Eduardo Rossel)

THE WORK OF THE FAO/ILO/IMO IN RELATION TO THE ISSUE OF SAFETY AND HEALTH FOR COMMERCIAL FISHERMEN

Andrew Smith, PhD (Open University)
Skippers Certificate of Competency, Teaching Certificate
Fishery Industry Officer
Fisheries Department
Food and Agriculture Organization, Rome

Background

It is estimated that 24,000 fatalities occur worldwide each year in fisheries. It seems plausible that the fatality rates in countries for which data are not available are higher than in those countries that do keep records. Recent reports from the Nordic countries indicate that fatality rates in fisheries range between 90 and 150 per 100,000, and yet the accident prevention, survival training, and search-and-rescue services offered in these countries are among the best in the world (The Working Party for Safety and Survival Training for Nordic Fishermen 2002).

From developing countries, much higher figures are cited. It has been estimated that fatality rates in Sri Lanka's offshore fisheries are ten times higher than in Norway (Dahl 1990); a study on fatality rates in canoe fishing in Guinea in 1991-1994 indicated a rate of 500 per 100,000; in a number of other countries along the West African coast, the artisanal canoe fatality rates appear to be in the range of 300 to 1,000 per 100,000 fishermen; and recent figures from South Africa report 585 fatalities per 100,000 fishermen (Food and Agriculture Organization 2000).

Iceland reported on recent findings concerning injuries in the Icelandic fisheries, which are generally regarded as highly mechanized and technically advanced. Considerable effort has been put into reducing the risk of these injuries in recent years. While there is reason to believe that fatal accidents in Iceland are fewer now than 10 years ago, the same is not the case with nonfatal injuries. Every year, 10% of all fishermen and 15% of fishermen on trawlers are subject to injuries. Accidents involving fishermen are more

3

common the longer they have been on the job, and there is a threefold risk of a fatal accident if the seaman has been more than 10 years on the job (Kristinsson 1999). Possible explanations are that the more experienced seamen are likely to be entrusted with the dangerous tasks or are more prone to taking risks. Also, younger crew members are more likely to have received safety training than older crew. This gives reason to hope that concerted efforts in improving safety education and training of fishermen, along with improved vessel design, construction, and working conditions on board, may result in reduced accident rates.

Improved safety at sea has for decades been of major concern to various institutions, national authorities, nongovernmental organizations, and individuals who recognize that a functional legal framework is the prerequisite for concerted actions for improved safety. The model for such legislation has already been provided by various international organizations.

UN Law of the Sea convention

The United Nations Conference on the Law of the Sea (hereafter referred to as the 1982 UN Convention) was completed in 1982 (United Nations 1982a), although its convention did not enter formally into force until 1994 when it had been ratified by the required number of states (for more information about nomenclature, see note a; for information about Exclusive Economic Zones [EEZs], see note b). The 1982 UN Convention had by October 2003 been ratified by 143 states. It is globally recognized as the regime dealing with all matters relating to the law of the sea and gives nations rights as well as responsibilities to utilize their living marine resources in a rational and sustainable way. Regarding safety, the 1982 UN Convention rules that every state shall effectively exercise its jurisdiction and control in administrative, technical, and social matters over ships flying its flag. Furthermore, the flag nation shall take such measures for ships flying its flag as are necessary to ensure safety at sea with regard to, inter alia, (a) the construction, equipment, and seaworthiness of ships (Petursdottir, Hannibalsson, and Turner 2001); the manning of ships, labour conditions, and the training of crews, taking into account the applicable international instruments; and (c) the use of signals, the maintenance of communications, and the prevention of collisions. In taking such measures, each state is required to conform to generally accepted international regulations, procedures, and practices and to take any steps necessary to secure their observance [Article 94(5)] (United Nations 1982b).

Other UN agencies

The International Maritime Organization (IMO), the International Labour Organization (ILO), and the Food and Agriculture Organization (FAO) are the three specialized agencies of the United Nations system that play a role in fishermen's safety at sea. IMO is the agency responsible for improving maritime safety and preventing pollution from ships. ILO formulates international labour standards in the form of conventions and recommendations, and sets minimum standards of basic labour rights. It also promotes the development of independent employers' and workers' organizations and provides training and advisory services to those organizations.

By virtue of their working methods, the results of IMO and ILO tend to have little impact on the safety of artisanal and small-scale fishermen. Most of the recommendations and conventions are aimed at large vessels, primarily the merchant fleet on international voyages. Some conventions explicitly exempt fishing vessels, and most do not apply to vessels under 24 meters long, thus leaving out the majority of fishing vessels and transport boats in developing countries.

International Maritime Organization and SOLAS

The first international convention concerning safety at sea was SOLAS (Safety of Life at Sea) and was prompted by the Titanic disaster in 1911. The convention was first adopted in 1914, with amendments adopted in 1929 and 1948. When IMO was founded in 1958, its first major task was amending SOLAS in 1960, and the organization has subsequently ensured that its revision is an ongoing process.

SOLAS specifies minimum standards for the construction, equipping, and operation of ships compatible with their safety. Apart from Chapter V, SOLAS does not apply to fishing vessels. Chapter V deals with safety of navigation and identifies certain navigation safety services that should be provided by contracting governments and sets forth provisions of an operational nature applicable in general to all ships on all voyages. This is in contrast to the convention as a whole, which only applies to certain classes of ships engaged in international voyages.

Recognizing that a number of fishing boat accidents are caused by submarines, a resolution was adopted by the IMO in 1987 recommending operational practices for submarines to reduce this danger.

Torremolinos convention and the Torremolinos protocol

The Torremolinos (Spain) International Convention for the Safety of Fishing Vessels, held in 1977, was the first-ever international convention on the safety of fishing vessels. It was intended to be a more formal document than the Code and Voluntary Guidelines (see below) and formulated more along the lines of the International Convention for Safety of Life at Sea in 1974. The convention contains safety requirements for the construction and equipment of new, decked, sea-going fishing vessels 24 meters long or more, including those vessels also processing their catch. Existing vessels were covered only with respect to radio requirements.

One of the most important features of the convention was that it contained stability requirements for the first time in an international convention. Other chapters dealt with such matters as construction, watertight integrity, and equipment; machinery, electrical installations, and unattended machinery spaces; fire protection, detection, extinction, and fire fighting; protection of the crew; lifesaving appliances; emergency procedures, musters, and drills; radiotelegraphy and radiotelephony; and shipborne navigational equipment.

Representatives of 45 countries agreed upon the convention in 1977, but subsequently it has not received sufficient ratifications to enter into force, as many states claim it to be either too stringent or too lenient for their fishing fleets. It was, therefore, decided to prepare a protocol to the convention. The purpose of the protocol is to overcome the constraints of the provisions in the parent convention that have caused difficulties for states and thereby enable the protocol to be brought into force as soon as possible (see note c). In several chapters, this was achieved by raising the lower size limit of vessels from 24 to 45 meters. The protocol also calls for the development of regional guidelines for vessels between 24 and 45 meters long, taking into account the mode of operation, sheltered nature, and climatic conditions of that region.

Standards of Training, Certification, and Watchkeeping for Fishing Vessel Personnel

The Standards of Training, Certification, and Watchkeeping for Fishing Vessel Personnel (STCW-F convention), which was adopted by IMO in 1995, contains requirements concerning skippers and watchkeepers on vessels 24 meters long and over, chief engineers and engineering officers on vessels of 750-kW propulsion power or more, and personnel in charge of radio communications. Chapter III of the Annex to the Convention includes requirements for basic safety training for all fishing vessel personnel. As of May 2000, the STCW-F Convention had been ratified by four countries.

Other related IMO conventions

Other IMO conventions that have particular relevance to safety and health in fishing include the International Convention on Maritime Search and Rescue, 1979, and the Convention on the International Regulations for Preventing Collisions at Sea (COLREGS), 1972 (as amended). Finally, the International Aeronautical and Maritime Search and Rescue Manual, whose purpose is to assist states in meeting search and rescue needs, contributes significantly to improving success rates in the rescue of fishermen.

This list of international conventions and recommendations shows that profound effort has already been invested at an international level in improving safety at sea. This work has been meticulously done, taking into account the design and construction of vessels, stability, load lines, mechanical equipment and gear, safety equipment, communications, effects of weather and icing, working conditions and hours, training of licensed personnel, etc. Thus, as has been repeatedly pointed out, there is no lack of regulations and administrative guidelines. What is missing is their effective enforcement at the national level.

International Labour Office

The prompt for the ratification of conventions for labour standards in the fishing industry has been a parallel, on-going process in the maritime industry. This process involves around 70 conventions and recommendations and has been termed the Super Convention in the discussion to rationalise all the maritime conventions. It was during the initial stages of these discussions–

when the employers' group (i.e., the International Shipping Federation) stated it was not competent to represent fishing industry employers–that it became evident that fishing industries would be deleted from several provisions of the maritime conventions and would not be included under the new convention (i.e., repatriations of seafarers and identity documents). This would have to be dealt with under another amalgamating convention for the fishing industry.

The present ILO conventions relating to the fishing industry include the following:

- Convention 112, Concerning the Minimum Age for Admission to Employment as Fishermen (1959),
- Convention 113, Concerning Medical Examinations for Fishermen (1959),
- Convention 114, Concerning the Vocational Training of Fishermen (1966),
- Convention 125, Concerning Fishermen's Certificates of Competency (1966), and
- Convention 126, Concerning Accommodation On Board Fishing Vessels (1966).

In addition, there are two recommendations:

- Recommendation 7, Concerning the Hours of Work in the Fishing Industry (1920) and
- Recommendation 126, Concerning the Vocational Training of Fishermen (1966).

If the parallel development of ILO conventions for the maritime industries is examined, a "trickle down" effect is obvious, in that as a convention was developed for the maritime industry, a similar convention was drafted for the fishing industry. This practice has failed in that a low number of ratifications has been secured for the fisheries conventions and recommendations. This may be due to the differences between the two industries–the maritime industry is characterised by a strong union-employee interaction, whereas the fishing industry is largely self-employed.

The current timetable for the ratification of the conventions included one preparatory conference in June 2004 and a final conference in June 2005.

It should, however, be noted that ILO has had strong support for more general conventions, including the establishment of a minimum age to prevent the worst forms of child labour. This convention has had a very high rate of ratification. Under this convention, employment is generally prohibited for persons under the age of 18 in certain occupations. Because the fishing industry is regarded as a hazardous industry, it would appear to fall under this category. However, there would appear to be enough flexibility in the convention to allow employment from the age of 16 (i.e., if the persons are adequately supervised and are not allowed to undertake hazardous tasks).

Regional arrangements

Some countries have included the issue of safety at sea in the work plans of regional bodies or organizations (such as the Organization of East Caribbean States [OECS]) (see note d), the Sub-Regional Fisheries Commission of North West African States (see note e), the South Pacific Commission (SPC) (see note f), and the Bay of Bengal Programme (BOBP) (see note g). In some cases, they have linked these work plans to fisheries management. Such arrangements will be of value during the formulation of standards intended to be adopted by all member countries through a programme for the harmonization of fisheries regulations.

Application of conventions and regulations to fisheries

At the national level, this same reason has hindered the inclusion of fishing vessels in regulations formulated by maritime administrations, while at the same time, industry representatives have, in some cases with success, lobbied for exemptions for a variety of reasons. This reflects reluctance on behalf of the fishing industry to be subjected to a comprehensive regulatory programme. Fisheries have a long tradition of independence. Many regard fisheries as the last frontier of free enterprise and resent government involvement, an involvement that may be perceived by the industry as inadequately informed of the risks and nature of fishing operations—or of the slim profit margins that might be eroded by mandatory compliance with regulations on training, vessel construction, and equipment. In addition, legislators may refrain from imposing laws or regulations on the fisheries that lead to additional costs or may otherwise be perceived as repressive. The US Coast Guard, for example, has repeatedly advocated the licensing and training of commercial fishing vessel crews, but to no avail. The US Congress has indeed drafted such legislation, but not enacted it into law. Research in the area of safety at sea for commercial fishermen in the United States has largely focused on the implementation and effectiveness of safety regulations. Findings strongly assert that fishermen's perceptions regarding safety can vary greatly from those of the government, including the Coast Guard, and that there needs to be a better understanding of the fishing culture and ways in which safety is viewed therein. These findings underscore the need to involve fishermen in the safety regulatory process; the "human factor" associated with safety at sea coupled with the cognitions and input of fishermen all provide essential information needed to make safety regulations more effective (Kaplan and Kite-Powell 2000).

Government policy to regulate for safety at sea in the fishing industry must be accompanied by a total commitment to implement that regulatory regime, along with the necessary resources. Implementation encompasses a set of strategies that could include education, assistance, persuasion, promotion, economic incentives, monitoring, enforcement, and sanctions, all of which are accompanied by setting up or improving administration and associated costs. Implementation must be considered at every phase of the regulation formulation—and not the final consequence of regulation.

While it may be true that "legislation is only as good as its enforcement," legislation cannot be improved by enforcement. The quality of the legislation remains the limiting factor. In many parts of the world, additional regulations for fisheries are not required. The overriding need is for regulations to be reviewed and amended to reflect problems and their root causes. The process of regulatory review must be as dynamic as the industry being regulated. Thus, it is clear the industry must be part of this process. The regulators and the regulated need the appropriate training to ensure compliance and enforcement, as well as a working relationship promoted by mutual respect and trust (Turner 2002). The establishment of national sea-safety working groups might be a step in the right direction. In some places the infrastructure necessary for enforcement hardly exists and would have to be built from scratch.

Food and Agriculture Organization (FAO): The Code of Conduct for Responsible Fisheries

The Code of Conduct for Responsible Fisheries was unanimously adopted by the FAO Conference in 1995. The code is voluntary. However, certain parts of it are based on relevant rules of international law, as reflected in the UN Convention on Law of the Sea of 10 December 1982. The code also contains provisions that may in the future be given–or have already been given–binding effect by means of other obligatory legal instruments among the parties, such as the Agreement to Promote Compliance with Conservation and Management Measures by Fishing Vessels on the High Seas, 1993. It is a unique instrument in its holistic approach, based on and bringing together key elements from the then-existing international conventions and guidelines concerning fisheries and related environmental issues (see note h). It offers guidelines for responsible fisheries, establishing principles and standards applicable to the conservation, management, and development of all fisheries. The code recognizes the nutritional, economic, social, environmental, and cultural importance of fisheries and the interests of all those concerned with the fishery sector. It also recognizes the importance of the safety issue and contains several separate references to the subject, addressing working and living conditions, health and safety standards, education and training, safety of fishing vessels, search and rescue, and accident reporting (see note i).

The fact that the code is to a great extent nonmandatory has proven to be more of an asset than a weakness. This renders it attractive as a model on which to base the management of fisheries, and its adoption does not carry the same formal consequences as the conventions it is based on. The code functions well as a model that can be applied under various conditions without the constraint of having to comply with standards that are not appropriate for the nation in question.

Every other year, FAO monitors to what extent the member states comply with the Code of Conduct. A response rate of 60% (during the year 2000) of all FAO member states, including landlocked countries, must be regarded as quite encouraging. Several countries have adapted the code to their fisheries and stages of developmentc, and it seems to serve well as a framework within which to build different types of management systems (Doulman 2003). It may be added that the Philippine Fisheries Code of 1998 closely follows the principles enshrined in the Code of Conduct. In addition to the code itself, FAO has prepared a series of Technical Guidelines for Responsible Fisheries, consisting at present of nine separate publications.

The Fisheries Department of FAO has been working in the field of fishermen's safety for 50 years, during which the fishing industry has been greatly affected by political, social, economic, and technological changes. These changes have led inexorably to increased pressure on fish resources. Consequently, governments have recognized that they need to be better aware of the state of their fisheries, to implement effective policies to prevent resource depletion and the wastage of fishery inputs, and, increasingly, to facilitate stock rehabilitation. While the extent and effect of fisheries management measures put in place around the world vary considerably, they tend to be more concerned with the long-term conservation and sustainable use of fisheries resources than with the welfare of those who harvest them. FAO, therefore, tends to concentrate on issues concerning the resources with only a few staff–people who generally have had sea-going experience and are interested in fisheries safety, but who can only devote a small proportion of their time to this very important subject.

Joint FAO/ILO/IMO work: Document for Guidance on the Training and Certification of Fishing Vessel Personnel

This Document for Guidance takes account of the conventions and recommendations adopted by ILO and IMO and of the wide practical experience of FAO in the field of training for fishing vessel personnel. It is intended to provide guidance when national training schemes and courses are instituted, amended, or developed for the vocational training of any category of fishing vessel personnel. It is stressed that additional guidance on training is complementary to, and not intended to supersede, the knowledge requirements specified in these ILO and IMO conventions and recommendations. The document applies to training and certification of both small-scale and industrial maritime fisheries. However, in the case of fishing vessels less than 24 m long or powered mainly by machinery of less than 750-kW propulsion power, certification is not prescribed, but may be introduced at the discretion of the competent administration. It is a revision of an earlier publication to take into account the STCW-F (1995), the FAO Code of Conduct for Responsible Fisheries, and recent developments in the fishing industry.

The FAO/ILO/IMO Code of Safety For Fishermen and Fishing Vessels and the FAO/ILO/IMO Voluntary Guidelines for the Design Construction and Equipment of Small Fishing Vessels

The Code of Safety comprises two parts: Part A (Safety and Health Practices for Skippers and Crews) and Part B (Safety and Health Requirements for the Construction and Equipment of Fishing Vessels). Both of these documents are currently under revision by an IMO intersessional correspondence group. In a similar fashion, the Voluntary Guidelines for the Design Construction and Equipment of Small Fishing Vessels is being revised.

National administrations

Maritime administrations, on the other hand, have the safety of seafarers among their prime concerns. However, they frequently have difficulty in adequately addressing the safety of fishermen because the nature of fishing operations is so different from that of the cargo handling and transport activities encountered in merchant shipping with which they are typically more familiar. Fishing vessels are excluded from the vast majority of provisions of international shipping conventions drawn up by bodies such as IMO, and to this day there is no international convention in force dealing with the safety of fishing vessels or the training of their crews.

The effect of this dilemma is that in many countries, the issue of fishing vessel safety is falling between the fisheries and maritime administrations and not being comprehensively addressed by either. Nevertheless, this problem has been recognized at an international level in the past, and a great deal of work has been done through a tripartite group of FAO/ILO/IMO. Unfortunately, the joint work of the three organizations does not trickle down to a national level.

Voluntary efforts

This is not to suggest that fishermen's safety can only be tackled through conventions, safety regulations, and enforcement. There are a number of areas in which improvements can be made: for example, more effective and holistic fisheries management; the training and certification of skippers and crews; increased collaboration between fishermen, their communities and organizations, and government; provision and analysis of data identifying the root and actual causes of accidents; and education and training of teachers, extensionists, and inspectors. All of these can lead to improvements. But the fact is that even in some of those countries where these measures are in progress, fishing-related deaths are on the increase. This is not because in those same countries there are more fishermen, or because their boats are less well built than they used to be—the reverse is the case. It is more likely to be the result of economic pressures and human factors such as risk taking, fatigue, stress, and simply attitude problems—a high risk of loss of life has been accepted as part of the fishing culture. This attitude has perhaps been one of the major underestimated obstacles to improved safety and work environment in the fishing industry and is one that can be corrected by concerted efforts from within the family and the community, as well as through government institutions. And indeed, it is important that it should. The consequences of loss of life fall heavily on the dependents.

In many developing countries, these consequences can be devastating. A widow may have a low social standing, and where there is no welfare state to support the family and without alternative sources of income, a widow and her children may face destitution.

The Code of Conduct for Responsible Fisheries notes that it is the duty of all states to "ensure that fishing facilities and equipment as well as all fisheries activities allow for safe, healthy, and fair working and living conditions and meet internationally agreed standards adopted by relevant international organizations."

How can states comply with this duty? Unsafe vessels could be eliminated. Safety equipment could be the subject of mandatory installation and servicing requirements. Awareness of stability issues could be achieved by improving crew training and education. Beach-based training on how to respond to emergencies at sea and how to use survival gear could be introduced for all crews. Administrations could reach out to fishing communities and fishermen's organizations to ensure their participaion and comprehensive involvement.

It is obvious that the above examples cannot be implemented overnight. But without them and others like them, any substantial improvement in safety at sea for current and future generations of fishermen is not achievable. Moreover, without them, fishermen will not be able to achieve a corresponding degree of social respect or status in acknowledgment of their membership in a professional group having recognized skills, expertise, standards, and responsibilities. Without that social respect or status, many small-scale and artisanal fishing communities will continue to exert little influence on society regarding the most fundamental issues that concern them, such as health, education, and infrastructure. In the absence of respect, these people will remain marginalized, left at the control of those who are able to exploit them.

To achieve progress and meaningful implementation of the types of measures described in the examples above, an enabling environment is a prerequisite. Sustainable results will not be achievable without political will, motivation, and commitment. In attempting to tackle the problem of fisheries safety, FAO has been trying to include safety at sea as an issue to be considered under fisheries management (Petursdottir, Hannibalsson, and Turner 2001). Often, too little attention is given to generating sufficient political support and lasting commitment for effecting the necessary social, technological, legal, and institutional changes needed to improve working conditions within the fisheries sector. It will be important to raise levels of awareness about sea safety issues among relevant senior government decision-makers.

The Code of Conduct contains many references to the obligations of countries concerning safety at sea. Such obligations notwithstanding, loss of life among fishing vessel crews continues to increase. While there is no lack of regulations and administrative guidelines at the international level, they are rarely enforced or implemented effectively at national levels. The promotion of responsible fisheries operations is especially problematical in the artisanal fisheries of developing countries. Although these fisheries account for the vast majority of the world's fishing fleets, they also reflect the environments where sea safety regimes are weakest.

References

Dahl EA (1990). Safety guidelines on design, construction, and operation of small offshore fishing boats in Sri Lanka, Bay of Bengal Project. Rome: Food and Agriculture Organization.

Doulman D (2003). Personal communication.

Food and Agriculture Organization (2000). The state of the world fisheries and aquaculture. Available at ftp://ftp.fao.org/docrep/fao/003/x8002e/x8002e00.pdf.

International Maritime Organization (1987). Avoidance by submerged submarines of fishing vessels and their fishing gear, Resolution A.599(15).

Kaplan M and Kite-Powell HL (2000). Safety at sea and fisheries management: Fishermen's attitudes and the need for co-management. *Marine Policy*.

Kristinsson S (1999). Note on the Proceedings of the Tripartite Meeting on Safety and Health in the Fishing Industry. Geneva: International Labour Organization.

Petursdottir G, Hannibalsson O, and Turner J (2001). Safety at sea as an integral part of fisheries management. Fisheries Circular 966. Rome: Fisheries Department, Food and Agriculture Organization.

Turner J (2002). Factors governing the development of national rules and regulations for the construction and equipment of small fishing vessels. *In* Proceedings of the International Fishing Industry Safety and Health Conference (Woods Hole, Massachusetts, Oct. 23-25, 2000), Lincoln JM, Hudson DS et al., eds. Cincinnati, OH: National Institute for Occupational Safety and Health. DHHS (NIOSH) Pub. No. 2002-147.

United Nations (1982a). United Nations Convention on the Law of the Sea of 10 December 1982.

United Nations (1982b). Article 94, Duties of the Flag State. Part VII: High Seas, Section 1: General Provisions.

Notes

a. Considerable confusion exists as to the proper use of the acronym UNCLOS with reference to the United Nations Conferences and Convention on the Law of the Sea. For a useful discussion on the topic see the *International Journal of Marine and Coastal Law*, Vol. 15, No. 3, and *Kluwer Law Journal International*, 2000, 12-07.
b. The clause on 200-mile EEZs had been agreed upon in 1976 with the effect that a number of nations extended their EEZs without delay.
c. The 1993 Protocol has been ratified by nine states.
d. Antigua and Barbuda, Commonwealth of Dominica, Grenada, Montserrat, St. Kitts and Nevis, St. Lucia, St. Vincent, and the Grenadines
e. Mauritania, Cap Verde, Senegal, the Gambia, Guinea Bissau, and Guinea.
f. Australia, Cook Islands, Federated States of Micronesia, Fiji, Kiribati, Nauru, New Zealand, Niue, Palau, Papua New Guinea, Republic of the Marshall Islands, Samoa, Solomon Islands, Tonga, Tuvalu, and Vanuatu.

g. Bangladesh, India, Indonesia, Malaysia, Maldives, Sri Lanka, and Thailand.

h. Certain parts of the Code are based on the 1982 UN Convention. It is to be interpreted and applied in conformity with the relevant rules of international law as reflected in the 1982 UN Convention, and in a manner consistent with the relevant provisions of the 1995 UN Fish Stocks Agreement, as well as in light of the 1992 Declaration of Cancun, the 1992 Rio Declaration on Environment and Development, and Agenda 21, especially Chapter 17. The FAO Compliance Agreement is an integral part of the Code. See Articles 1 and 3 of the Code.

i. Reference is made to issues directly pertaining to safety in paragraphs 6.17: 8.1.5: 8.1.6: 8.1.7, 8.1.8, 8.2.5: 8.3.2: and 8.4.1

Session Two

SESSION TWO:
LESSONS LEARNED: FINDINGS FROM RECENT INVESTIGATIONS OF A COMMERCIAL FISHING VESSEL MISHAP

Boats in eastern Russia *(Photo courtesy of Alan Sorum)*

THE *ARCTIC ROSE*: FORENSIC ANALYSIS OF A CASUALTY

James B. Robertson, Lieutenant Commander
U.S. Coast Guard, Marine Safety Office Anchorage
Chief, Inspections Department
Recorder, Marine Board of Investigation
Anchorage, Alaska, USA
E-mail: jrobertson@cgalaska.uscg.mil

The purpose of this paper is to explain the findings of the Marine Board of Investigation in its search for the proximate cause of the sinking of the fishing vessel *Arctic Rose* and to provide details and discuss the forensic analysis techniques employed. Errors discovered during the investigation in the SafetyNET system are identified Finally, key recommendations made to Commandant as a result of the investigation's findings are discussed.

Methodology: The use of modern technology and a novel dynamic stability model developed by Coast Guard naval architects are discussed to demonstrate the characteristics of the *Arctic Rose* at the time of the casualty. This stability concept allowed the Coast Guard to narrow the scope of its investigation and focus its energies on improving commercial fishing vessel safety.

Results: Through detailed analyses of all facts and testimony, the Coast Guard was able to determine the proximate cause of the sinking of the *Arctic Rose* and the events leading up to its demise.

Conclusions and Recommendations: Several safety recommendations for the commercial fishing vessel industry are evaluated with a focus on stability-related issues and standardization of stability operating tables.

The *Arctic Rose* was a 92-foot fishing vessel, western-rigged with the pilot-house forward and trawl deck aft with a large A-frame gantry at the stern. The ship operated primarily in the Gulf of Alaska and the Bering Sea as a head-and-gut processor, that is, the catch was deheaded and gutted in preparation for freezing and packaging in the processing space.

On April 2, 2001, the *Arctic Rose* encountered tragedy on the high seas, sinking between 2200 April 1 and 0335 April 2 (all times are Alaska Standard Time). The Seventeenth Coast Guard District Command Center received a 406 EPIRB (electronic position-indicating radio beacon) alert at 0335, issued an Urgent Marine Information Broadcast (UMIB), and sent an IN-MARSAT C message to all stations to alert other vessels. A search-and-rescue case was initiated, and Coast Guard aircraft were sent to the position of the EPIRB. At 0840, a Coast Guard C-130 arrived on the scene and located the vessel's EBIRB at 58°56.9'N/175°56.3'W. A large debris field and oil sheen were found in the vicinity. A nearby fishing vessel eventually responded to the UMIB and joined the search-and-rescue efforts. Shortly after arriving on-scene, the *Alaskan Rose* recovered the body of David Rundall from the water. A subsequent search by Coast Guard aircraft, two cutters, and two samaritan fishing vessels in the immediate area failed to recover additional personnel. Fourteen persons are missing at sea and are presumed dead. There were no survivors.

The Coast Guard's Marine Safety Office in Anchorage, Alaska, immediately began investigating the casualty. The Commandant of the Coast Guard recognized the significance of the casualty and convened a Marine Board of Investigation, the Coast Guard's highest level of investigatory body. Once the Marine Board members were selected, each member began researching the vessel's history and investigating all leads. The Marine Board faced a daunting task of trying to re-create and reconstruct the *Arctic Rose* at the time of her sinking. This case presented the investigators with several roadblocks. First, the investigators had no witnesses or survivors of the sinking. Second, the *Arctic Rose* was a one-of-a-kind vessel, built by backyard boat builders without plans. Finally, shifts in the commercial fishing industry in the Seattle, Washington, region displaced many of the vendors and technicians who had provided services to Arctic Sole Seafoods during refurbishment of the *Arctic Rose*.

The investigation was jointly conducted by investigators from the Coast Guard and the National Transportation Safety Board in various locales in

Alaska, Washington, and other western states. The investigators interviewed over a hundred witnesses in preparation for the hearing to gather facts concerning the *Arctic Rose* and its operation. The hearings for the Marine Board were held in two locations (Anchorage and Seattle) to accommodate witness travel and reduce costs. The Board received testimony from 55 witnesses addressing diverse topics as vessel stability, vessel operations, manning, industry practices, weather conditions, communications, and the Coast Guard's response to the accident.

The Marine Board launched an expedition to locate the wreck of the *Arctic Rose* in order to conduct an underwater survey of the vessel as it searched for more possible clues and answers to the mystery. The Board felt it was important to find the wreck of the *Arctic Rose* and conduct a survey of the vessel in search of clues to help identify the proximate cause of the casualty. The Board chairman petitioned Coast Guard Headquarters in Washington, DC, for permission and funding to carry out an expedition to locate and conduct underwater surveys of the vessel through the use of a remote operated vehicle (ROV). Coast Guard Headquarters granted the Marine Board's request, and two expeditions were ultimately organized to locate, survey, and videotape the condition the wreck of the *Arctic Rose*. Various contractors were organized to provide a suitable vessel to use as a platform for the ROVs and their support teams. The Marine Board was fortunate to locate a vessel, the M/V *Ocean Explorer*, already outfitted with a sonar array under charter to National Marine Fisheries Service and ready for deployment. The use of ROVs to locate and survey the wreck of the *Arctic Rose* was a critical tool for the Board. Two pieces of equipment were used during the first expedition, a Klein 5000 sonar array and a Phantom submersible.

The Klein 5000 is an extremely sensitive side-scanning sonar with the ability to detect minute objects or details on the sea floor. The Klein 5000 system consists of a towfish, tow cable, transceiver/processor unit, and a personal computer for system control and data viewing. The stainless steel towfish incorporates two multi-channel acoustic arrays and a pressure bottle, which houses all the electronics and sensors necessary for sonar recording, altitude sensing, system control, and telemetry. The sonar and sensor data are transmitted up the tow cable via a high-speed digital telemetry link, requiring only a single co-axial or fiber-optic cable. The surface-mounted transceiver/processor unit receives the data, performs all necessary digital processing functions on these data, and relays control commands to the towfish. It requires a team of two for its operation, one technician to fly the "fish" underwater and the other to monitor the sonar picture. The Klein 5000 system and her crew proved to be critical in locating the *Arctic Rose*. The equipment for the sonar array was located in a small space just forward of the engine room of the M/V *Ocean Explorer*. This space became the operations center for the search.

The voyage from Unalaska, Alaska, to the search area took approximately 2 days. During that time, the expedition team readied their equipment and developed a comprehensive search plan. The Marine Board provided known locations of the composite EPIRB hit, debris field, oil slick, and liferaft to the sonar team, who used the information to build a search grid and search pattern. The team integrated technology of a Triton Elics International Isis Sonar digital acquisition system with a side-scan sonar Klein 5000 towfish. Data from the Klein 5000 was transmitted up 300 meters of tow cable.

Once on-scene, the M/V *Ocean Explorer* conducted a pass on the initial trackline using its bottom-scanning sonar to check for any possible snags that might entangle or damage the Klein 5000. During the third sonar pass, a large target was found. Subsequent passes revealed a silhouette that matched the profile of the *Arctic Rose*. Using these techniques, the team located the wreckage and identified the *Arctic Rose* soon after the start of the search. The *Arctic Rose* lies at a depth of 428 feet 200 miles northwest of St. Paul Island in the Bering Sea. Several additional passes were made, producing high-resolution images of the wreck and scanning for any debris that might have entrapped the ROV.

The following morning, the ROV Phantom HD2 was readied, lowered into the water, and operationally tested. The ROV motored along the bottom with its video camera sending pictures to the surface (these images were recorded for the Marine Board). A hull came into view, and the ROV went alongside the hull, rising as it traveled. Finally, letters came into view, confirming that the *Arctic Rose* was located. The ROV was at the port bow of the vessel and proceeded toward the pilothouse. The video showed that the vessel was resting upright on the sea floor with a slight starboard list. The ROV attempted to power toward the stern of the vessel, but became hopelessly tangled in loose net-mending twine and was lost when the umbilical parted in a last-ditch attempt to free it. The Marine Board received approximately 14 minutes of usable video.

The Klein 5000 was placed back into service, and the *Ocean Explorer* made several close passes of the wreck, hoping to obtain any additional clues about the sinking of the vessel. The expedition reluctantly returned to Unalaska, knowing its mission was unfinished.

The Marine Board received permission from Coast Guard Headquarters to prepare a second expedition to survey the wreck of the *Arctic Rose*. On August 20, this second expedition departed Unalaska aboard the M/V *Ocean Explorer* with a larger, more-powerful ROV to return to the *Arctic Rose* and complete its mission. The MAXRover was equipped with stronger thrusters and a robotic arm that could be used to cut the ROV free from debris.

The team arrived on-scene several days later, but was forced to loiter due to rough sea conditions. Finally the weather calmed to within the minimum weather window for safe ROV operations. The ROV was lowered over the side and operationally tested. The MAXRover descended to the wreck. The video camera was activated and filmed the wreckage for clues. The MAXRover completed five dives and surveys of the wreckage. The Marine Board was able to examine the entire starboard hull, stern/transom area, trawl deck along with all associated equipment, and the exterior of the processing space. The Marine Board was also able to view the aft port section of the hull, keel cooler, shaft, kort nozzle, and rudder. A great deal was learned about the vessel and its condition.

The use of the ROV and sonar equipment allowed the Marine Board to gain some first-hand knowledge of the condition of the *Arctic Rose*, which proved critical to the Board in reaching its conclusion as to the cause of the sinking.

The Marine Board discovered the following details concerning the vessel:

- No hull failure or excessive corrosion was seen.
- No damage or indication that the vessel had been rammed or had struck an object prior to the sinking was observed.
- No buckled decking or side shell insets were noted.
- The vessel's rudder was hard over to port.
- The vessel appeared to have struck the bottom stern first.
- The aft weathertight door to the processing space was open. A starboard guillotine closure for the by-catch overboard discharge chute was open.
- Heavy gear was strewn across the deck and resting over the starboard bulwarks of the vessel.

+ The vessel's trawl doors were missing.
+ The vessel's trawl net was on the net reels.
+ The vessel's propeller and shaft were in place.

The discovery of these facts allowed the Marine Board to discount or eliminate many theories as to why the vessel sank, but raised a few questions, especially concerning the missing trawl doors. However, the facts provided more answers than questions and assisted the Marine Board in arriving at the most probable cause of the casualty.

SafetyNET

SafetyNET is an internationally adopted, semi-automated satellite service designed for the promulgation of maritime safety information to all types of vessels. SafetyNET broadcasts are made over the INMARSAT-C system of geostationary satellites and are free of charge. Virtually all navigable waters of the world are covered by INMARSAT satellites. It is a part of the global maritime distress and safety system (GMDSS). This system provides for automatic distress alerts in cases where a radio operator does not have time to send an SOS or Mayday call and, for the first time, requires ships to receive broadcasts of maritime safety information. If the INMARSAT-C satellite terminal is connected to a global positioning satellite or similar navigational receiver, or the ship's position has been recently updated manually, the vessel's position will be transmitted as part of the automatic distress alert. Specific radio carriage requirements depend on the ship's area of operation rather than its tonnage. The system also provides redundant means of distress alerts and emergency sources of power.

The designation of priorities in the SafetyNET system determines the order in which a message is broadcast. INMARSAT-C is a store-and-forward system where messages of higher priority are placed at the head of the queue for broadcast. The two highest priorities, distress and urgent, also set off the alarms of certain shipboard INMARSAT-C terminals, notifying the mariner that a high priority message had been received.

Most INMARSAT-C terminals will not receive a safety broadcast if it is transmitting a message or if it is tuned to an INMARSAT ocean region not used for safety broadcasts in the area traveled. Most SafetyNET messages are rebroadcast after 6 minutes to give a transmitting terminal time to receive missed messages. Lists of SafetyNET broadcast schedules and areas have been published by the World Meteorological Organization to assist ship operators in tuning INMARSAT-C terminals to the proper INMARSAT ocean region.

The Seventeenth District Command Center in Juneau issued a SafetyNET broadcast at 0429 on April 2, relaying the *Arctic Rose* distress information with a service parameter of "navigational warning" and a priority parameter of "distress." Because of the configuration of Telenor's system prior to November 2001, the message defaulted to "safety priority" based on the service parameter. No documentation provided to the SafetyNET users at rescue coordination centers indicated that the priority of messages would be determined by the service parameter, and they were unaware of the system's ability to default a message to a lower priority. Testimony provided to the Marine Board substantiated that Coast Guard users were unaware of the system's default settings. Coast Guard personnel from TISCOM, Headquarters, and COMSAT held a series of meetings in which the software problem was identified and corrected. Telenor's system was modified in November 2001 so that a message with any service parameter could be broadcast with any priority. Finally, the Coast Guard implemented training for all rescue coordination center personnel on the proper use and formatting of SafetyNET message configuration to avoid any unnecessary message delays in the future.

Stability

The Marine Board requested technical assistance from the Coast Guard's Marine Safety Center (MSC) in conducting an independent stability analysis to determine the most likely cause of the loss of the *Arctic Rose*. The stability calculations were performed using Creative Systems' General Hydrostatics (GHS) Version 7.50 software. MSC evaluated 19 different scenarios that could have led to the loss of the *Arctic Rose*. MSC used the best estimate of the loading condition of the vessel at the time of the casualty as the baseline for all stability calculations.

Based on these loading conditions, at the time of the casualty the *Arctic Rose* would have met the righting arm characteristic criteria and severe wind and roll criteria listed in Navigation and Vessel Inspection Circular (NVIC) 5-86. This assumes that the processing space had been maintained completely weathertight as required by the stability letter from Jensen Maritime Consultants' (JMC) dated July 9, 1999.

Dr. Bruce Johnson, chair of the Society of Naval Architects and Marine Engineers Ad Hoc Panel on Fishing Vessel Operations and Safety, worked in concert with Lieutenant George Borlase of MSC to develop a progressive flooding analysis spreadsheet. This forensic analysis tool is based on quasi-static time steps through various progressive flooding scenarios into as many as six interior compartments where large free-surface effects would have negatively affected the vessel's stability.

The analysis established the three most likely causes of progressive flooding into the processing space: (1) from a wash-up hose left on or from the water supply to the plate freezers, (2) from the aft deck through the open aft door by boarding seas, and (3) through the open aft door if the vessel took a roll to starboard of only 23°. Regardless of how the water entered the processing space, subsequent stability would have been very reduced, and progressive flooding would have continued until the vessel sank. Had the processing space been maintained as weathertight as per the JMC stability booklet, the *Arctic Rose* would not have sunk.

The loss of the *Arctic Rose* was most likely caused by progressive flooding from the aft deck into the processing space through the door in the aft bulk-head of the processing space. Flooding probably continued rather rapidly forward through the open door in the forward bulkhead of the processing space. The water then flooded the galley and engine room through nonwatertight doors. Initial flooding of the lazarette/dry stores space or progressive flooding into the machinery space was not necessary for the vessel to lose all positive righting arm. In fact, progressive flooding of the processing space and fish hold alone would have caused the vessel to lose all positive righting arm due to the large free-surface effect. Flooding of these two spaces alone also suggests a much slower net flooding rate and therefore a much longer time to sink. The vessel would very likely have lost all positive stability between 90 seconds and 4 minutes after progressive flooding started and to have sunk in as little as 3 minutes once progressive flooding began.

The arrangement of the *Arctic Rose* increased the likelihood of progressive flooding from the processing space. The door from the processing space to the aft deck was far outboard on the starboard side, reducing the heel angle at which water could enter the processing space. In addition, the doors leading forward into the galley and the engine room were also on the starboard side of the vessel. A lolling angle to starboard caused by the inflow of water through the aft door and the free-surface effect inside the processing space would have caused water to spill forward easily into the galley, down into the engine room, and eventually into the fish hold. The fish hold had a center-line hatch and would not flood significantly until enough water was in the processing space to spill into the fish hold. Also, it was very likely that none of the doors or hatches at the main deck were properly closed, which would have increased the likelihood of progressive flooding throughout the ship.

The *Arctic Rose* had a stability test conducted in 1991, after its conversion from the *Sea Power*, which generated a set of operating conditions for the vessel with the restriction that "These stability calculations assume the processing area is intact and watertight. If water accumulates in the processing area all fishing or processing operations must be halted until the water is cleared." In 1999, after the *Arctic Rose* was purchased by Mr. David Olney, JMC was hired to generate a new stability booklet for the vessel. On March 31, 1999, a new inclining experiment was conducted to calculate the light-ship displacement and centers of gravity for the vessel. Operating limits for the vessel were then created based on the stability criteria found in NVIC 5-86, entitled, "Voluntary Standards for U.S. Uninspected Commercial Fishing Vessels." The stability booklet was signed on July 9, 1999, and contained restrictions on freeboard, tank usage, and the amount of cargo carried on deck and underdeck. The operating instruction's second paragraph stated that, "This stability letter is void unless the processing space is kept weather-tight at all times."

In Section IX, "Weather Tightness and Seaworthiness," the operating instructions require that "All watertight doors shall be kept closed except when used for passage...Doors for the scrap chutes and the fish chutes in the factory bulkheads should be kept closed at all times except when necessary to conduct processing operations. All side fittings that open to the factory must be fitted with a watertight closure and check valve."

A large number of weight additions, removals, and relocations were performed on the *Arctic Rose* between July 9, 1999, when the operating instructions were issued, and April 2, 2002. The stability calculations performed by JMC accounted for a 13,500-pound keel shoe (ballast bar) at the time of the inclining experiment. The owner later added approximately 20,000 pounds of a boiler shot-cement mixture (boiler shot is a term used to describe round steel pieces approximately 1-2 inches in diameter), which was poured into the shaft alley area of the fish hold of the vessel after the inclining. In addition, a plate freezer and new refrigeration equipment were added in the processing space, a water maker was installed, and other equipment was added to the vessel. The owner did not track any of the weight additions, re-locations, or removals for the vessel. Compounding this error, the owner did not contact a naval architect to evaluate the effect of the weight changes on the vessel's stability. Furthermore, the owner testified to his belief of meeting the operating chart through the addition of weight with the keel shoe and boiler shot-cement mixture. In fact, this was in error as these weights were accounted for in the JMC stability calculations.

The *Arctic Rose* was not in compliance with the operating instructions issued by JMC at the time of the casualty. The aft starboard door in the processing space was open and the guillotine closure for the starboard discharge chute was two-thirds open, preventing the processing space from being weather-tight. The fuel and water tanks were being used in the opposite order speci-fied in the stability letter. A review of the *Arctic Rose* stability booklet and testimony provided to the Marine Board indicated the consumption order of the wing tanks had a negligible effect on the vessel's stability. The double-bottomed fuel tank was not kept pressed full at all times, but was instead being used as a day tank. Testimony provided to the Marine Board from a former chief engineer indicated the double-bottomed fuel-oil tank was used as a day tank and was refilled at the beginning of each day. There was between 9,500 and 12,000 gallons of fuel oil on board the vessel, and 53,000 pounds of product, stores, and ballast stored in the fish hold at the time of the casualty. According to the deck loading table from the JMC stability booklet, maximum deck load (which included both processing and cod end-loads) was 3,000 pounds. However, there was 10,000 pounds of deck load in the plate freezers at the time of the casualty. While independent calcula-tions later found the vessel met the intact stability criteria, at the time of the accident, the master of the vessel had use only of the operating instructions to evaluate whether his vessel met the minimum stability criteria.

Recommendations

The Marine Board made over 20 safety recommendations to the Comman-dant. These recommendations were divided into several categories, including regulatory changes, policy, and training.

1. The Coast Guard should develop regulations in which all watertight and weathertight doors that are required to be closed by a vesel's stabil-ity booklet to be alarmed and equipped with a visual and audible system

in the pilothouse to indicate the position of the door(s). If the aft door
to the processing space, which was required to be closed at all times by
the *Arctic Rose's* stability booklet, had been equipped with such an alarm
on the bridge, which would have sounded until the door was properly
secured, the sinking of the *Arctic Rose* could have been prevented. There
is a strong likelihood that future casualties of this nature could be prevented if this recommendation is implemented industry-wide.

2. In reviewing the overall SafetyNET system, the Marine Board found no
requirements for the use of an INMARSAT-C system on fishing vessels. This reduces the effectiveness of an important link of the GMDSS.
The Marine Board recommended requiring all fishing vessels operating beyond the boundary line to be GMDSS compliant. The Marine
Board understands that "one size fits all" requirements may not be the
right solution and further recommends that the Commandant evaluate
the possibility of a regulation based on a regional approach and tied to
vessel operations, number of persons on board, duration of voyage, and
distance offshore. The Federal Communications Commission (FCC)
and Coast Guard should partner during the development of these
regulations. Finally, the FCC and Coast Guard should require each fishing vessel equipped with a GMDSS system to have a properly trained
operator. There are two types of INMARSAT-C systems sold for use
aboard ship: a GMDSS version and a non-GMDSS version, commonly
referred to as the fisheries version. The two versions are very similar and
provide many of the same features. However, with the non-GMDSS
version, messages and safety broadcasts are often received and stored
internally, without any notification to the operator that a message has
been received. Although reception of SafeyNET traffic is automatic, the
shipboard operator must set the proper parameters on the receiver at the
start of the voyage. This includes the following steps:

   + Select the appropriate broadcast channel. This can often be accomplished by logging on to the land earth station in the ocean region
from which needed broadcasts are made.
   + Select the NAVAREA indentification code.
   + If traveling near Australia, select the proper coastal area codes.
   + Ensure that the INMARSAT-C station is connected to a working navigational receiver. If a connection cannot be made, the ship's
position must be manually updated every 4 hours during the ship's
voyage. Without these updates, countless unnecessary broadcast
messages will be received.

3. The INMARSAT-C system aboard the *Arctic Rose* did not have an audible or visual alarm to notify the watchstander of an incoming urgent broadcast. The user would have had to go from the steering station to the INMARSAT-C unit and download messages. Each message would have to be viewed prior to deleting it from the queue. The system operator has to program the INMARSAT-C system to receive messages based on the location of the terminal to avoid overloading the system with messages from other broadcast stations. The *Alaskan Rose's* mate provided testimony to the Marine Board indicating the vessel did receive countless messages from Russia, but did not receive the distress message until several hours after it had been sent by the Coast Guard. This was due to the system not being properly configured. It is imperative to "program" the INMARSAT-C system properly to receive messages. Although reception of SafetyNET traffic is automatic, the shipboard operator must set up the receiver properly at the start of the voyage. The most critical step is to ensure that the INMARSAT-C station is connected to a working navigational receiver. If a connection cannot be made, the ship's position must be manually updated every 4 hours during the ship's voyage. Without these updates, countless unnecessary broadcast messages will be received.

4. The Marine Board received testimony from a naval architect and member of the Society of Naval Architects and Marine Engineers' (SNAME) Ad Hoc Panel on Fishing Vessel Operations and Safety Working Group B indicating that the average commercial fisherman is not familiar with stability information. Furthermore, stability information is provided in a myriad of forms as there is no set industry standard. This creates an environment where stability information is presented to the mariner in a format that can be difficult to read and/or interpret. As a result many fishermen determine the stability of their vessels by feel.

5. The information on the *Arctic Rose* is open to wide interpretation. A previous mate on the *Arctic Rose* reviewed the JMC stability booklet and stated, "I think if I had this information and I had seen this particular stability book, I would not have gone on the *Arctic Rose*."

6. The Coast Guard should encourage the use of color graphic displays within a stability booklet that are easily understood by mariners, such as the one under development by SNAME . The work group is promoting a format that presents stability information and operating guidelines in a color graphic that is easy to understand. This format provides the vessel's operator and crew with a quick visual reference to make an informed decision for safe operations without having to perform stability calcula-

tions. This format was presented to several witnesses during the Marine Board hearings and was well received by the fishermen, especially when placed side by side with other graphs, which were open to interpretation. The color graphic displays provide a quick visual reference and allowed each person shown the visuals to make a quick go/no-go decision.

7. The Coast Guard should encourage the development of fishing vessel construction standards that minimize the free flow of water through a vessel. In addition, the Coast Guard should remove all provisions that allow the use of above-main-deck spaces in the development of a fishing vessel's stability characteristics. This two-fold recommendation stems from the construction of the *Arctic Rose*. During its conversion from a shimp/scalloper trawler to a head-and-gut processing trawler, the fish processing space was added at the main deck level.

8. The Coast Guard and the commercial fishing industry should explore the development of a minimal safety indoctrination program for all first-time crew. Such training would include processors prior to getting underway. A means to document the training should also be found. This recommendation would expand existing regulations and provide all people working on vessels at sea with a basic training course so that all participants would have an overview of fishing vessel operations and the proper use of safety equipment aboard a vessel. Upon successful completion of the course, a participant would receive wallet-sized card that could be presented to vessel owners/operators when filling out a job application as proof of completion of the indoctrination course.

In closing, the purpose of the Marine Board and any Coast Guard investigation is not to place blame, but to determine the cause of the event and make appropriate safety recommendations to prevent future occurrences. The sinking of the *Arctic Rose* was a tragic accident and has affected countless lives forever. How the *Arctic Rose* sank will never truly be known. The use of modern technology and forensic naval architecture combined with superb investigative work aided the Marine Board in providing answers to many questions and reach sound conclusions.

Session Three

SESSION THREE:
RISK FACTORS FOR COMMERCIAL FISHERMEN–
A LOOK AT THE DATA

Bringing home the catch in Pakistan (Photo courtesy of Dr. Muhammad Khan)

NONFATAL INJURY SURVEILLANCE AND PREVENTION IN THE BERING SEA CRAB FISHERY

Bradley J. Husberg, MSPH, BSN
Jennifer M. Lincoln, MS
Centers for Disease Control and Prevention
National Institute for Occupational Safety and Health
Alaska Field Station,
4230 University Drive
Anchorage, Alaska, USA

Introduction

Commercial fishing is one of the most dangerous occupations in the United States. In 2002, commercial fishermen had the second highest traumatic injury fatality rate of all workers in the United States—71.1/100,000 workers, which is 16 times the national rate of 4.4/100,000 workers across all industries (Bureau of Labor Statistics 2003). Only timber cutters had a higher fatality rate of 117.8/100,000 (Bureau of Labor Statistics 2003). Many fishermen work in harsh environments with high winds, cold water, and icy conditions. The deck of a fishing vessel is a congested work area, crowded with fishing gear and equipment, and is constantly moving. The National Research Council noted "the apparent high incidence of workplace accidents suggests inadequately designed safety features in machinery, deck layouts, and fishing gear" (National Research Council 1991). Other factors that could impair safety in this workforce include isolated locations, long working hours, and days with little rest. Many commercial fishermen endure fatigue, physical stress, and financial pressures to push their vessels and crew to make their livings (Lincoln and Conway 1997; Conway, Lincoln et al. 2002).

Commercial fishing fatalities in Alaska are a result of the vessel sinking, falls overboard, or severe injuries from machinery and fishing gear on deck. Extensive research has been conducted in describing the fatality problem in the Alaskan fishing industry and progress in saving lives after a vessel has been lost (Lincoln and Conway 1997, 1999; Lincoln, Husberg, and Conway 2001). Many (41%) nonfatal injuries in Alaska are due to machinery and fishing equipment (Thomas, Lincoln et al. 2001; Husberg, Lincoln et al.

2001). Conclusions from these papers include the statement that "further efforts are required to redesign or install safety features on fishing machinery and equipment." Various fleets and fisheries use their own specific fishing equipment and fishing methods, so tailored techniques are needed.

It is challenging to look at nonfatal work-related injuries in any industry. Problems arise when defining what a nonfatal injury is. Many times the completeness and accuracy of the reporting system is questioned. In the commercial fishing industry, the US Coast Guard has the following requirements for reporting nonfatal injuries:

> 46CFR4.05: An injury that requires professional medical treatment (treatment beyond first aid) and, if the person is engaged or employed on board a vessel in commercial service, that renders the individual unfit to perform his or her routine duties...

However, this reporting requirement is not regularly enforced, and no mechanism is in place to measure how accurately these injuries are being reported. Fortunately, Alaska has another surveillance system to collect data on work-related nonfatal injuries—the Alaska Trauma Registry (ATR). The state maintains the ATR as a quality assurance tool for the state's trauma system. It has allowed identification of causes and specific information for useful injury surveillance in the fishing industry (Husberg, Lincoln et al. 2001; Thomas, Lincoln et al. 2001).

The ATR is a population-based trauma registry administered by the State of Alaska, Department of Health and Social Services, Division of Public Health, Section of Community Health and Emergency Medical Services. Technical assistance and funding for work-related injury surveillance are provided by the National Institute for Occupational Safety and Health (NIOSH), Division of Safety Research, Alaska Field Station. To be included in ATR, patients must have sustained a traumatic injury as identified through a review of their hospital discharge diagnosis (ICD-9-CM code 800.00 through 995.89). Work-related cases in the ATR meet NIOSH Operational Guidelines for Determination of Injury at Work (Marsh and Layne 2001). Information is collected for all patients admitted to any of the 24 hospitals in Alaska who have been injured severely enough to be hospitalized for more than 1 day.

In October 2000, the NIOSH Alaska Field Station partnered with engineers and naval architects at Jensen Maritime Consultants, Inc. (JMC), to develop injury prevention recommendations to address deck hazards in the commercial fishing industry. Here, we describe the nonfatal injuries recorded in the ATR from 1991-1999 and the initial steps in developing the "Injury Prevention in the Commercial Fishing Industry" project.

Methods

Injury surveillance data from ATR from 1991-1999 were used to identify causal factors and circumstances surrounding work-related injuries in the Alaskan commercial fishing industry. We categorized the results into cause of injury, types of injury, and the body region injured. We also reviewed the injury narratives and identified the Bering sea crab fishery as the fleet to use to start our deck safety efforts. After consolidating the injury data, project staff met with groups of crab fishermen in Seattle, Washington, and Kodiak, Alaska, to get input on the possible causes and potential injury prevention solutions to injuries. The feedback from these sessions was used to create a survey to gather further information and comments regarding potential solutions to deck safety. Staff from the Coast Guard and JMC administered the survey in Unalaska, Alaska, during the crab fleet's pre-season vessel inspections in October 2001. Survey results were completed, and interventions were identified from these results.

Results

From 1991-1999, 39,143 injuries were recorded in ATR. Approximately 10% (3,951) were classified as work-related, and 648 of these were in the commercial fishing industry.

Data from ATR show that the top three primary causes of commercial fishing injuries were machinery (205), followed by falls (163), and struck by object (100). The three most common types of injury were fractured bones (309), open wounds (77), and burns (32). The body regions most commonly injured were the upper extremities (203), lower extremities (189), and head (96). A review of the narrative field for machinery identified crab pot launcher, crane, and bait chopper as the most frequently cited pieces of machinery involved in injuries.

The Coast Guard and JMC surveyed 89 fishermen on 75 vessels. Information from these surveys helped identify priority areas and create recommendations for the improvement of safety on these vessels.

Through the process of using ATR surveillance data, working with fishermen, conducting surveys, and analysis of the resultant information, several topics were generated that could lead to the development of recommendations. These included—

+ Improving deck visibility
+ Installing closed-circuit television
+ Installing mirrors
+ Improving lighting
+ Painting moving deck machinery in bright colors
+ Working more safely with crab pots
+ Installing pot guides
+ Identifying and marking launcher's danger zones
+ Developing means of preventing slips and falls
+ Covering manholes
+ Installing nonskid gratings and mats
+ Studying the best rail heights
+ Installing seawalls
+ Developing procedures for man-overboard recovery
+ Modifying other miscellaneous equipment
+ Installing bait chopper guards
+ Improving crane maintenance and index markings
+ Developing hydraulics maintenance and emergency procedures

The specific topics noted above are discussed in the paper by Blumhagen et al. in these proceedings and are further explained in the "Deck Safety for Crab Fishermen" booklet published by JMC (2002). This booklet also contains information on the cost of implementing the recommendations.

Conclusion

Surveillance information from the ATR has assisted in identifying and prioritizing causes of nonfatal injuries on the deck of crab-fishing vessels. This first phase has led the way to specific injury prevention recommendations. The second phase is currently focusing on fisheries other than Bering Sea crab fishing in Alaskan waters. Collaboration between NIOSH, JMC,

US Coast Guard, University of Utah School of Engineering, and the fishing industry is continuing to identify and recommend specific injury prevention measures for this dangerous industry.

References

Conway GA, Lincoln JM et al. (2002). Surveillance and prevention of occupational injuries in Alaska: A decade of progress, 1990-1999. Cincinnati, OH: National Institute for Occupational Safety and Health. DHHS (NIOSH) Pub. No. 2002-115.

Husberg BJ, Lincoln JM et al. (2001). On-deck dangers in the Alaskan commercial fishing industry. Marine Safety Council.

Jensen Maritime Consultants, Inc. (Seattle, Washington) (2002). Deck safety for crab fishermen (information booklet).

Lincoln JM and Conway GA (1997). Commercial fishing fatalities in Alaska: Risk factors and prevention strategies. Cincinnati, OH: National Institute for Occupational Safety and Health. Current Intelligence Bulletin 58. DHHS (NIOSH) Pub. No. 97-163.

Lincoln JM and Conway GA (1999). Preventing commercial fishing deaths in Alaska. *Occupational and Environmental Medicine* 56: 691-695.

Lincoln JM, Husberg BJ, and Conway GA (2001). Improving safety in the Alaskan commercial fishing industry. *International Journal of Circumpolar Health* 60(4): 705-713.

Marsh SM and Layne LA (2001). Fatal injuries to civilian workers in the United States, 1980-1995–national and state profiles. Appendix V. Cincinnati, OH: National Institute for Occupational Safety and Health. DHHS (NIOSH) Pub. No. 2001-129.

National Research Council (1991). Fishing vessel safety: Blueprint for a national program. Washington, DC: National Academy Press.

Thomas TK, Lincoln JM et al. (2001). Is it safe on deck? Fatal and non-fatal workplace injuries among Alaskan commercial fishermen. *American Journal of Industrial Medicine* 40: 693-702.

Bureau of Labor Statistics (2003). Washington, DC: U.S. Department of Labor.

SESSION FOUR:
DECK SAFETY

Faroese fishermen at work (Photo courtesy of Anna Maria Simonsen)

Session Four

USE OF OPERATING HAZARD ANALYSIS TO REVIEW ON-DECK PROCEDURES IN COMMERCIAL CRAB FISHING

Donald S. Bloswick, PhD, PE, CPE
University of Utah
Department of Mechanical Engineering
50 S. Central Campus Drive
Salt Lake City, Utah, USA
E-mail: bloswick@mech.utah.edu

Bradley J. Husberg, MSPH, BSN
Centers for Disease Control and Prevention
National Institute for Occupational Safety and Health
Division of Safety Research, Alaska Field Station
Anchorage, Alaska, USA

Eric Blumhagen, PE
Jensen Maritime Consultants
Seattle, Washington, USA

Introduction

Information from 1991-1998 indicates that commercial fishing in Alaska has an occupational fatality rate approximately 28 times the rate for US workers in general (Thomas, Lincoln et al. 2001). Lincoln notes that within the commerical fishing industry, the Alaska shellfish fishery has the highest fatality rate, which is approximately twice as high as the rate for herring, the fishery with the next highest rate.

Until recently, when surpassed by construction, commercial fishing has also resulted in the largest number of work-related injuries in Alaska industry (Husberg, Lincoln et al. 2001). In 1983, injury rates for the Dungeness, tanner, and king crab fisheries were among the highest of all fisheries (Bender 1992). Being struck by crab pots has been found to be the single most common cause of injury due to equipment (Thomas, Lincoln et al. 2001).

One result of the Second National Fishing Industry Safety and Health Workshop held in 1997 was the recommendation to "perform job hazard analysis on those tasks associated with increased injuries" (Klatt and Conway 2000). Tomasson (2002) proposed a review of all work procedures on board ships to reveal which work procedures were hazardous. Thomas et al. (2001) suggested that efforts are required to "better define the relationship between the vessel, fishing equipment, and the fishermen" and noted that while data were lacking on the human aspect of the system, "additional strategies to improve safety need to address the interaction between the vessel, its equipment and machinery, and the worker." Husberg, Lincoln, et al. (2001) emphasized that there was a need to examine the deck environment from a "mechanical and safety engineering perspective." They also noted the use of "cranes, 'power blocks,' pulleys, winches, lines, nets, crab pots, and crab pot launchers" is an issue requiring attention.

Operating hazard analysis

This paper uses operating (and support) hazard analysis (OHA) to analyze systematically the job hazards in several on-deck tasks in commercial crab fishing. Examples from cod fishing with pots are also used. OHA is a systems safety technique often used to (1) describe and quantify (to the extent possible) the hazard associated with processes that are inherently dangerous or in which human error is likely to cause injury or property damage and (2) provide recommended risk reduction alternatives during all phases of tasks or operations. OHA concentrates on the performance of people and their relationships to potential task hazards. For a particular on-deck operation, severity and probability can be quantified (to some extent) through the use of epidemiological and historical data and from estimates by knowledgeable personnel. The OHA procedure presented here is a modification of that presented by Vincoli (1993). It should be noted that in this paper, "hazard" is defined as a condition with the potential to cause injury or property damage. The existence of a hazard does not imply an inevitable result. The factors appearing frequently in marine casualty literature are listed below (National Research Council, Marine Board Committee on Fishing Vessel Safety 1991).

+ Fatigue/stress.
+ Improper or inadequate procedures (including inadequate or unsafe loading and stability practices and inadequate watchkeeping).
+ Improper maintenance.
+ Inattention (including carelessness).

+ Inadequate human engineering in design.
+ Inadequate physical condition.
+ Incapacitation through use of alcohol and drugs.
+ Inexperience (including inadequate knowledge and skills and insufficient familiarity with the vessel or fishing activity).
+ Judgmental errors (including risk-taking and faulty decision-making).
+ Navigational/operator error (including inexperience and errors in judgment).
+ Neglect (including willful negligence).
+ Personnel relationships.
+ Working conditions.

The intent of this paper is to provide a starting point for the use of systems safety procedures to analyze systematically on-deck commercial fishing operations. The commercial crab fisheries are featured in this study. The methods may be used in other fisheries making up the commercial fishing industry. Abatement recommendations are presented; however, some are the result of previous studies and are not the primary intent of this paper.

A form (Figure 1) to facilitate the application of OHA is illustrated. This form is an expansion of one presented by Vincoli (1993).

The initial entries on the form are a simple list of each procedure or task (with an identifying number) and a description of the potentially hazardous conditions in the task. The cause (if known) of this hazardous condition and possible effect are next noted. Completion of the form also requires an estimate of the frequency(Table 1), severity (Table 2), and detectability (possibility that the hazardous condition will be detected before it results in an adverse event) (Table 3) for each condition. A general measure of concern for the potentially hazardous condition can be estimated by combining frequency, severity, and detectability.

Frequency

The hazard frequency levels presented in Table 1 are based on those included in MIL-STD 882B (US Department of Defense 1984). The frequency levels represent a qualitative judgment of the likelihood that a mishap will occur if the hazard is not corrected.

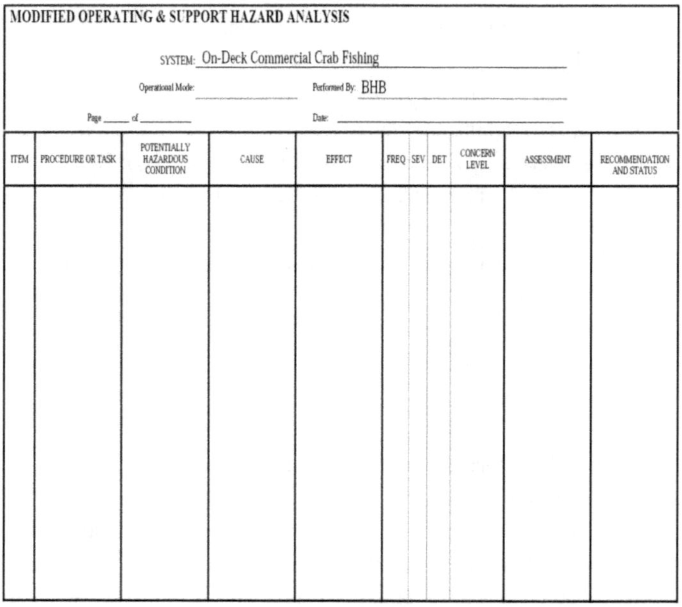

MODIFIED OPERATING & SUPPORT HAZARD ANALYSIS										
SYSTEM: On-Deck Commercial Crab Fishing										
Operational Mode:			Performed By: BHB							
Page ___ of _____			Date:							
ITEM	PROCEDURE OR TASK	POTENTIALLY HAZARDOUS CONDITION	CAUSE	EFFECT	FREQ	SEV	DET	CONCERN LEVEL	ASSESSMENT	RECOMMENDATION AND STATUS

Figure 1: Form for operating and support hazard analysis

Table 1: Hazard probability categories

Description	Level	Definition
Remote	1	Unlikely, but may possibly occur in the life of an item
Occasional	2	Likely to occur sometime in the life of an item
Probable	3	Will occur several times during the life of an item
Frequent	4	Likely to occur frequently

Severity

In a similar fashion, hazard severity levels are presented in Table 2 (US Department of Defense 1984). The severity levels represent a qualitative judgment of the relative severities of the outcome of the uncorrected hazard.

Table 2: Hazard severity categories

Description	Category	Definition
Negligible	1	Less than minor injury, occupational illness, or system damage
Marginal	2	Minor injury, occupational illness, or system damage
Critical	3	Severe injury, occupational illness, or system damage
Catastrophic	4	Death or system loss

Detectability

The detectability metric is intended to reflect the probability that a hazard will be detected and corrected before it results in a mishap. The use of this metric assumes that detected hazards will be corrected, which is not necessarily a correct assumption. (A measure of detectability is sometimes included as part of the judgment of frequency or the likelihood that a mishap will result from a particular hazard.) Table 3 contains possible measures of detectability.

Table 3: Hazard detectability categories

Description	Level	Definition
Easy	1	Hazard obvious. Knowledge of hazard is "second nature.
Moderate	2	Hazard can be detected with usual effort. Operators generally aware of hazard.
Difficult	3	Hazard can be detected with unusual effort. Operators may be aware of hazard.
Improbable	4	Existence of hazard usually not detected. Operators generally unaware or unconcerned about hazard.

The evaluation of detectability is sometimes a critical component when evaluating the hazards of mechanical systems. In the case of on-deck crab fishing operations, this metric will also include a measure of the operator's likely awareness of the hazard. Frequency, severity, and detectability are combined to represent a concern level. The concern level is a general measure of how resources should be allocated to get the most "bang for the buck." The combination of frequency, severity, and detectability may be qualitative or quantitative through addition or multiplication of the three scores. In this paper, the three metrics are multiplied to accentuate potentially high-hazard procedures. The assessment column provides for any needed discussion of the concern level and the recommendation and status column allows for summary recommendations to be proposed and abatement status to be recorded.

General procedures in crab fishing

On-deck procedures during crab fishing, or other fishing operations utilizing pots, vary somewhat depending on the size and design of the boat on which the operations take place. However, several things must happen, regardless of boat size, that relate directly to the fishing operation. For example, the pots must be loaded onto the boat, moved to the launcher, prepared for launching, launched, retrieved, emptied, moved to on-deck storage, and unloaded from the boat. In addition, bait must be prepared, and the catch must be off-loaded from the hold. Support activities, such as loading supplies on board, general boat maintenance and repair, and galley operations, are not included. A sequential list of the general procedures in crab fishing is given below.

1. Load/off-load crab pots to/from on-deck storage; stack and secure on boat.
2. Move pots from on-deck storage stack to launcher.
3. Prepare pot for initial launching, coil line, bait pot, secure door.
4. Launch pot and throw out line and buoys.
5. Retrieve buoys, connect line to power block, bring in pot.
6. Attach pot to picking boom crane and move pot to launcher.
7. Move sorting table, transfer catch from pot to table.
8. Sort and move crab to hold.
9. Prepare pot for repeat launching (approximately the same as procedure 3) and repeat steps 4-8.
10. Move pots from launcher to on-deck storage.

Major items not in the above sequence include chopping bait and general on-deck movement. These procedures are illustrated in Figures 2 through 12 by number. Estimates of frequency, severity, and detectability are based on entries into ATR for 1991-1998 (n = 80) and the best initial estimates by a small group of professionals with experience in commercial on-deck crab fishing activities. It is anticipated and desirable that the identification of potentially hazardous conditions, as well as estimates of frequency, severity, and detectability, will be modified with further review of existing epidemiological data, the use of additional years of trauma registry data, and additional input from interested parties. The concern levels are intended to represent a relative ranking of the hazardous conditions in the included procedures and tasks. As noted earlier, the intent of this paper is to provide a starting point for the use of systems safety procedures to analyze on-deck commercial crab (or other) fishing operations systematically. While draft recommendations are sometimes presented, they are generally the result of previous studies and intended to stimulate discussion. They are not intended to be the primary result of this paper.

References

Bender TR (1994). Commercial fishing fatalities: US regional comparisons. *In* Proceedings of the National Fishing Industry Safety and Health Workshop (Anchorage, Alaska, Oct. 9-11, 1992), Myers ML and Klatt ML, eds. Cincinnati, OH: National Institute for Occupational Safety and Health. DHHS (NIOSH) Pub. No. 94-109.

Husberg BJ, Lincoln JM et al. (2001). On-deck dangers in the Alaskan commercial fishing industry. Marine Safety Council.

Lincoln JM, Husberg, BJ, and Conway GA (2001). Improving safety in the Alaskan commercial fishing industry. *Internat. Journal of Circumpolar Health* 60(4):705-713.

Klatt ML and Conway GA (2000). Working group recommendations. *In* Proceedings of the Second National Fishing Industry Safety and Health Workshop (Seattle, Washington, Nov. 21-22, 1997), Klatt ML and Conway GA, eds. Cincinnati, OH: National Institute for Occupational Safety and Health. DHHS (NIOSH) Pub. No. 2000-104.

National Research Council (1991). Fishing vessel safety: Blueprint for a national program. Washington, DC: National Academy Press.

Thomas TK, Lincoln JM et al. (2001). Is it safe on deck? Fatal and non-fatal workplace injuries among Alaskan commercial fishermen. *American Journal of Industrial Medicine* 40: 693-702.

Tomasson G (2002). Safety management on board Icelandic fishing vessels. *In* Proceedings of the International Fishing Industry Safety and Health Conference (Woods Hole, Massachusetts, Oct. 23-25, 2000), Lincoln JM, Hudson DS et al., eds. Cincinnati, OH: National Institute for Occupational Safety and Health. DHHS (NIOSH) Pub. No. 2002-147.

Vincoli JW (1993). *Basic Guide to System Safety*. New York: Van Nostrand Reinhold.

Figure 2 : Load/off-load crab pots

Figure 3. Move pots from on-deck storage stack to launcher

Figure 4: Prepare pot for initial launching

Figure 5: Lauch pot and throw out line and buoys

Figure 6: Retrieve buoys, connect line to power pot, bring in pot

Figure 7: Attach pot to picking boom crane and move pot to launcher

Figure 8: Move sorting table, transfer catch from pots to table

Figure 9: Move and sort crab to hold

Figure 10: Move pots from launcher to on-deck storage

Figure 11: Chop bait

Figure 12: General on-deck movement

Deck Safety
Crab deck safety hazards

	PROCEDURE OR TASK	POTENTIALLY HAZARDOUS CONDITIONS	CAUSE	EFFECT	FREQ.	SEV	DETECT.	CONCERN	RECOMMENDATIONS
1	Load/unload pots to/from boat	High hand forces and awkward postures	Push/pull pots Climb on pots	Musculoskeletal injury/illness	1	2	2	4	Worker Training
		Moving pots/caught between	Pots move without worker knowledge Pots move too fast Pots shift due to boat movement	Crushing injury	2.5	3	2	15	Worker Training Increase communication between worker and crane operator/captain Improve control layout on crane More secure tie-off to dock
		Fall from height	Work above deck on stack of pots Slippery work surface	Contact trauma	1	4	1	4	Worker Training Footwear PPE?
		Fall on same level	Slippery work surface	Contact trauma	1	2	2	4	Worker Training Footwear
		Pot fall from overhead	Crane/gear failure Crane operator failure	Contact trauma	1	4	3	12	Crane maintenance Operator training Improve control layout on crane
2	Move pots from on-deck storage to launcher	High hand forces and awkward postures	Push/pull pots	Musculoskeletal injury/illness	1	2	3	6	Worker Training
		Caught between pot or other obstruction or struck by pot	Pot moves without worker knowledge Pot moves too fast Pot swings due to boat movement	Crushing/contact trauma injury	2.5	3	2	15	Worker Training Increase communication between worker and crane operator/captain Improve control layout on crane
		Pinch point at feet	Front edge of launcher moves toward deck	Foot caught between launcher and deck, crushing injury	1	3	2	6	Worker Training Identify pinch point area with bright paint Launcher "feet" or "pads" on deck (stumble hazard?)
		Fall from height	Work above deck Slippery work surface	Contact trauma	1	4	1	4	Worker Training Footwear PPE?
		Fall on same level	Slippery work surface	Contact trauma	1	2	2	4	Worker Training Footwear
		Pot fall from overhead	Crane/gear failure Crane operator failure	Contact trauma	1	4	3	12	Crane maintenance Operator training Improve control layout on crane
3	Prepare pot for initial launching	High hand forces and awkward postures	Open/close pot door	Musculoskeletal injury/illness	1	2	2	4	Worker Training
		High hand forces and awkward postures	Remove buoys and rope Hang bait	Musculoskeletal injury/illness	1	1	2	2	Worker Training

PROCEDURE OR TASK	POTENTIALLY HAZARDOUS CONDITIONS	CAUSE	EFFECT	FREQ	SEV	DETECT	CONCERN LEVEL	RECOMMENDATIONS
4 Launch pot and throw out rope and buoys	Awkward postures while throwing	Bulky items	Musculoskeletal injury/illness	1	1	3	3	Worker Training
	Struck by launcher	Launcher movement	Contact trauma	1	1	1	1	N/A
	Moving rope	Uncoiling rope overboard	Man overboard	1	4	2	8	Training Warning sound PPE
5 Throw line, retrieve buoys, connect line to power block, bring in pot	Contact with hook	Hook movement	Contact trauma	1	1	1	1	N/A
	Awkward dynamic movement during throw	Hook weight Throw distance	Musculoskeletal stress at shoulder	1	1	3	3	
	Pull line in, high hand forces and awkward (hands at/above shoulder) postures		Musculoskeletal stress at shoulder	1	1	2	2	Worker Training
	Power block, caught in in-running nip point	Hand too close to in-running nip point Loose clothing	Crushing injury	1	3	2	6	Worker Training Tape sleeves Guarding?
6 Attach pot to picking boom and move pot to launcher, unhook from picking boom	Caught between pot and launcher or other obstruction	Pot moves without worker knowledge Pot moves too fast Pot swings due to boat movement Launcher tilts to deck	Crushing/contact trauma injury	1	3	2	6	Worker Training Increase communication between worker and crane operator/captain Improve control layout on crane
	Pinch point at feet	Front edge of launcher moves toward deck	Foot caught between launcher and deck	1	2	2	4	Worker Training Identify pinch point area with bright paint Launcher "feet" or "pads" on deck (stumble hazard?)
7 Relocate sorting table, transfer catch to table	Caught between table and launcher or other obstruction	Table moves without worker knowledge	Crushing/contact trauma injury	1	2	1	2	Worker Training
	Awkward posture and high repetition during catch transfer	Location of table with respect to pot	Musculoskeletal stress	1	1	2	2	Worker Training
8 Sort and move catch to hold	Awkward posture, high repetition, torso flexion	Need for speed, location of table with respect to hold	Musculoskeletal stress to shoulder and back	1	2	1	2	Worker Training
	Contact with sharp objects (crab, cod)	Catch characteristic, need for speed	Puncture wound	4	1	1	4	PPE (Gloves)
	Contact with sharp objects (crab, cod)	Catch characteristic, need for speed	Eye injury	1	3	1	3	PPE (Eye protection)

	PROCEDURE OR TASK	POTENTIALLY HAZARDOUS CONDITIONS	CAUSE	EFFECT	FREQ.	SEV	DETECT	CONCERN LEVEL	RECOMMENDATIONS
9	Move pots from launcher to on-deck storage	High hand forces and awkward postures	Push/pull pots	Musculoskeletal injury/illness	1	2	3	6	Worker Training
		Caught between pot and other obstruction	Pot moves without worker knowledge Pot moves too fast Pot swings due to boat movement Launcher tilts to deck	Crushing/contact trauma injury	2.5	3	2	15	Worker Training Increase communication between worker and crane operator/captain Improve control layout on crane
		Fall on same level	Slippery work surface	Contact trauma	1	2	2	4	Worker Training Footwear
		Fall from height	Work above deck Slippery work surface	Contact trauma	1	4	1	4	Worker Training Footwear PPE?
		Pot fall from overhead	Crane/gear failure Crane operator failure	Contact trauma	1	4	3	12	Crane maintenance Operator training Improve control layout on crane
10	Chopping bait - mechanical	Nip point	Blades on bait chopper	Injury to hand	2	4	1	8	Worker Training Guarding Design of bait chopper to minimize perceived need to get hand close
	Chopping bait – manual (axe)	Movement of sharp object	Axe	Injury to hand	1	3	2	6	Worker Training Covers over small hatches Guardrails around large hatches
11	General on-deck activities	Fall on same level	Slippery work surface	Contact trauma	1	2	2	4	Worker Training Footwear
		Wash or fall overboard	High waves Boat movement	Hypothermia Drowning	2	4	1	8	Worker Training PPE Rescue procedures
		Contact with structure and equipment	High waves Boat movement	Contact trauma	1	2	1	2	Worker Training PPE

PRACTICAL DECK SAFETY FOR CRAB FISHERS

Eric Blumhagen, PE
Jensen Maritime Consultants, Inc.
4241 21st Avenue West
Seattle, Washington, USA

Bradley J. Husberg, MSPH, BSN
Centers for Disease Control and Prevention
National Institute for Occupational Safety and Health
Division of Safety Research, Alaska Field Station
4230 University Drive
Anchorage, Alaska, USA

Purpose

Crab fishers face some of the highest occupational injury and death rates in the nation. In Alaska, the fatality rate for the shellfish fishery was higher than all other fisheries in the industry with 356 fatalities per 100,000 fishers per year between 1991 and 1996, or about 50 times the national average (Lincoln and Conway 1997). While most fatalities were caused by vessels sinking or person-overboard events, most of the nonfatal injuries were caused by deck machinery or falls on board the vessel. The goal of this study was to find practical, inexpensive solutions to deck safety problems and disseminate that information to fishers. We approached the problem of deck safety from the fishers' perspective, using extensive input from fishers during the course of the project.

This project was not intended to be a basis for any type of new regulation for deck safety installations. It was only intended to provide fishers with the information required to make improvements on their boats if the modification were appropriate for their particular arrangements and circumstances. The nonregulatory approach was critical to the success of the project. Without this approach, we would have had great difficulty in getting ideas from fishers. We would also have had far more limited fisher participation in the survey.

Methods

The initial analysis of the most common injuries on crab fishing boats was taken from the Alaska Trauma Registry (ATR), which records all traumatic injuries requiring hospitalization in Alaska (Lincoln, Husberg, and Conway 2002). By its nature, this database only includes the more severe injuries. The ATR includes a short narrative description of the circumstances surrounding the injury. A sample description from the ATR reads "left hand caught in a bait grinding machine on board vessel." These descriptions were used to categorize incidents into a particular fishery, as well as sort them by type of incident (slip/fall, machinery, etc.). This research helped determine where the hazards are on deck, allowing us to better focus our efforts during the rest of the project.

The next step was to spend time on board a crab boat during fishing operations, both to observe likely hazards and to document the process of setting, retrieving, and otherwise handling pots. The use of all deck equipment was observed and documented in photos and video recordings. Project staff spent 2 days on board the *Royal Viking* out of Akutan, Alaska, observing pot cod fishing, which uses the same equipment and procedures as crab fishing.

Armed with first-hand knowledge of fishing operations, project staff held a focus group meeting with a small group of crab boat skippers, each of whom had over 20 years of experience in the Alaska crab fisheries. We discussed modifications they had already made to their boats to reduce deck injuries, toured one of the boats to view the modifications, and discussed which safety issues were most likely to be addressed by physical changes to the vessels.

After the focus group meeting, we developed a list of proposed modifications to vessels, combining the ideas listed in the focus group meeting with some of our own ideas. These ideas were used in a survey to be given in port to the largest group of crab fishers possible immediately prior to a crab season. We asked the fishers if each idea or modification would help improve deck safety and whether it had been implemented on their boat. If the modification was not in place, we asked why it had not been incorporated. We also collected basic information such as number of years experience in the fishery, vessel name, and crew position. Where necessary, we inserted drawings or photos into the survey to help explain ideas to fishers.

The survey was administered during October 2001 in Dutch Harbor, Alaska, immediately before the start of the Bristol Bay red king crab season. A total of 89 fishers from 75 different boats were surveyed by project staff and US Coast Guard personnel. Forty-one of the fishers surveyed were skippers, while the others held other positions in the crew. Forty-four of those surveyed had more than 20 years of experience in the fisheries, while 11 had less than 5 years experience. We believe that the survey sample reflects a representative cross-section of crab fishers.

Results

Following are interventions discussed in the survey.

Group I: Visibility

1. Install adequate lighting on deck. This increases visibility in dark areas and during night and helps the crew work more safely (Figure 1).
2. Use a closed-circuit TV system on house-forward boats. This helps the skipper see what is happening on deck from the wheelhouse. He or she can then become more easily aware of dangerous conditions and activity on deck (Figure 1).

Figure 1: Closed-circuit TV camera (in the round white housing) and extra lighting at bait chopper

Figure 2: Truck mirror seen from inside wheelhouse

3. Install a truck mirror on the starboard side of the wheelhouse. This helps the skipper see work along the starboard rail and pot launcher (Figure 2).

Group II: Machinery

4. Install lock valves on cranes, haulers, or winches on older machinery. These valves make the machinery hold the load when the hydraulic valve is in the neutral position. On most hydraulic machinery, this is not a problem, since most machinery already has these valves installed.

5. "Footprint" the pot launcher. A small half-round or half-oval strip around the deck where the pot launcher meets the deck will outline hazardous areas. This will help the crew feel when their feet are starting to get into a dangerous working zone (potentially leading to crushed feet, toes, or lower extremities).

6. Install pressure relief valves on the pot launcher. This would keep the launcher from crushing a person underneath the launcher when it is being lowered from the upright position.

7. Install an emergency shut-off for the launcher near the launcher. This allows the crew to shut down the launcher quickly if a person is caught while it is moving.

8. Paint the hazardous zone around the launcher, and/or the launcher itself, a contrasting color. This helps the crew see the danger area around the pot launcher and makes the moving part of the launcher more visible.

9. Install "pot guides" on the outside of the bulwark (Figure 3). Pot guides are triangular stops that run vertically on the outside of the

Figure 3: Pot guides

bulwarks. In heavy seas, pot guides help the crew control the crab pot after it is out of the water, but before it is pulled over the rail on to the launcher.

10. Install a guard over the bait chopper. This helps prevent the crew from reaching into the chopper box and risking injury (Figure 4).

11. Install an emergency shut-off on the bait chopper. This allows another person to turn off the bait chopper if a fisher either has been injured by the chopper or is about to be injured.

Figure 4: Bait chopper guard made from old conveyor belt material.　　*Figure 5: Boat with raised bulwark on port side and raised pipe rail on starboard side*

12. Mark the crane to help align it with the pot launcher. This allows the crew to position the crane head rapidly and accurately over the pot launcher, reducing the chance of injury from a swinging pot as it is being lifted or lowered at the launcher.

Group III: Crew protection

13. Install a raised bulwark on the port side or around the entire working deck. This provides a sheltered area that helps protect the crew from large waves coming over the side (Figure 5).

14. Install gratings over hold manholes. This prevents a fall into the hold manhole if the watertight cover must be removed. If the watertight cover is removed, some of the water circulating in the hold flows out on deck, helping reduce ice buildup on the deck.

15. Increase the rail height along the perimeter of the working deck. This helps prevent a wave from washing a crew member overboard (Figure 5).

16. Install nonskid grating in low-wear areas. Most crab boats are fitted with a steel or wood grating above the actual deck to prevent the crab pots from wearing away the watertight deck. Nonskid grating provides better footing in low-wear areas. However, it cannot be installed everywhere, since it will not stand up to the wear from the crab pots.

17. Install man-overboard recovery devices (life rings, life slings, flares, etc.) at the stern of the boat or at the hauling station. This helps the crew respond quickly to a person-overboard event.

After the data from the survey had been analyzed, we created a booklet describing each of the ideas we discussed in the survey. Since no idea received less than a 25% overall "approval rating," the project team decided to include all the suggested interventions in the booklet. However, some items were consolidated into general sections.

The project team also added some ideas to the booklet based on the surveys and discussions with fishers. The most significant of these was a section on emergency preparedness. Several fishers told us about the importance of planning ahead for emergencies, creating procedures for responding, and drilling the crew on those procedures. The emergency most commonly mentioned was person overboard. In this case, fishers said it was important to practice both returning the vessel to the person in the water and retrieving the person from the water.

The booklet gives the fishers all the information required to make an informed decision about whether each safety idea will work in given situations. For each idea, the booklet has a description of the idea with drawings and photos as necessary, what problem it is intended to solve, and how to go about making the change to the vessel. Approximate costs for implementing each idea were also included. The booklet was published and has been distributed widely throughout the West Coast, with particular emphasis on Alaska.

To further illustrate the improvement ideas, a model of a typical crab boat showing most of the safety ideas was made to display at conventions and trade shows. The model has attracted quite a bit of discussion and interest wherever it has been displayed.

Conclusions

This project has been an effective tool in finding low-cost, effective solutions to deck safety problems on commercial crab vessels and in distributing the information to the fleet. Fishers have been very willing to participate in the project and have appreciated the information presented in the booklet and deck model. Much of the success of this project has been due to the non-regulatory approach and assistance from fishermen's associations and the US Coast Guard. The success and acceptance of this project has led to a similar project working with smaller vessels engaged in longline, troll, seine, dive, and small-pot crab fisheries.

References

Lincoln JM and Conway GA (1997). Commercial fishing fatalities in Alaska: Risk factors and prevention strategies. Cincinnati, OH: National Institute for Occupational Safety and Health. Current Intelligence Bulletin 58. DHHS (NIOSH) Pub. No. 97-163.

Lincoln JM, Husberg BJ, and Conway GA (2002). Improving safety in the Alaska commercial fishing industry. In Proceedings of the International Fishing Industry Safety and Health Conference (Woods Hole, Massachusetts, Oct. 23-25, 2000), Lincoln JM, Hudson DS et al., eds. Cincinnati, OH: National Institute for Occupational Safety and Health. DHHS (NIOSH) Pub. No. 2002-147.

AN EXPERIMENTAL COMPARISON OF MARINE VESSEL DECK MATERIALS AND FOOTWEAR SLIP RESISTANCE UNDER VARYING ENVIRONMENTAL CONDITIONS

Brad Wallace, MS, CSP
Ponchatoula, Louisiana, USA

Introduction

Commercial fishing continues to rank at or near the top of the most hazardous occupations in the United States (US Coast Guard 1999). Commercial fishing is a complex industrial process that that can vary greatly among fisheries (NIOSH 1994). This type of work is often performed in extreme weather conditions and on unstable platforms. Work may be performed at extremes of temperature, daylight, and work hours. According to 1995 statistics, commercial fishing ranked as the most hazardous occupation in America. Second that year were sailors and deckhands on other classes of vessels (US Coast Guard 1999).

Historically, fishing vessel safety has been an on-going struggle between the rights of fiercely independent individuals willing or resigned to accept the hazards of their profession, and those from within and outside the industry who attempt to mitigate the extreme dangers of the business (US Coast Guard 1999). Numerous initiatives have been proposed and rejected or dampened. Most legislation enacted focuses on preparation and survival of catastrophic accidents and the availability of safety equipment during and after a disaster at sea. Prevention measures have been recommended by numerous organizations, but few have been developed into enforceable standards for the commercial fishing industry. Governmental agencies that have made occupational safety recommendations directed at the commercial fishing industry include the National Institute for Occupational Safety and Health (NIOSH), the Occupational Health and Safety Administration, the Coast Guard, the National Transportation Safety Board, the National

Marine Fisheries Service, and the National Research Council, to name a few. The almost unanimous conclusion from these organizations was to direct occupational safety initiatives that focus on accident prevention rather than incident mitigation.

Over 21% of the fatalities occurring in Alaskan waters involve falls on deck or falls overboard (Lincoln and Conway 1997). In 1994, these two types of falls were identified as major hazards aboard commercial fishing vessels and recommended the use of safety lines, guard rails, and nonskid decking materials. In 1997, this suggestion was reemphasized when NIOSH recommended that all fishermen wear personal flotation devices (PFDs) while on deck surfaces (Linoln and Conway 1997). The same report listed prevention efforts as critical factors that must be addressed; however, the use of PFDs was emphasized rather than the prevention of falls in the first place. The Coast Guard's landmark report from the Fishing Vessel Casualty Task Force, *Living to Fish, Dying to Fish* (1999), also identified falls as a common hazard on fishing vessels and recommended several items to be addressed relating to falls. These items include better research and development in the area of human factors engineering, occupational safety awareness, industrial safety standards, wearing of PFDs, and the development of "good marine practices."

In preventing slips and falls, the concept is simple: Use a flooring (deck) surface that has a coefficient of friction high enough not to be "slippery" (English 1996). This is can be accomplished in a number of ways; however, in the marine industry, it is frequently addressed by the application of "nonskid" deck coatings, particularly on steel-decked vessels. In a commercial fishing environment, decks can become more slippery because of water, ice, snow, oils, fish/shellfish body tissues and fluids, and other potential deck contaminates. Friction cohesion can also be altered by deck angles and motions caused by sea conditions.

The performance and effectiveness of deck materials and coatings in combination with personnel footwear remain relatively unevaluated as they relate to personnel falls in the commercial fishing industry. The problem to be dealt with in this study was to analyze and compare the effectiveness of various available deck materials and coatings and footwear under laboratory conditions that would simulate the actual environmental conditions that might be encountered during commercial fishing activities. There was also a need to validate the marketing claims of those producing an increasing variety of deck surfaces and footwear available to vessel owners, builders, and operators.

In land-based industrial settings, slip resistance of floor surfaces may be generally measured through the use of tribometers or horizontal pull dynamometers to calculate a coefficient of friction and apply it toward a standard. Based on consensus standards, proposed regulations, and case law (English 2000), the standard for a safe level of friction is generally recognized to be a coefficient of friction of 0.50 or higher. It should be noted that this is for level surfaces. The Americans with Disabilities Act recommends wheelchair access ramps have a minimal static coefficient of friction of 0.80 or higher (English 2000).

Slip resistance is a complicated science. It is generally recognized among researchers that slip resistance is affected by a number of factors, including shoe bottom material and texture, floor (deck) material and finish, environmental contaminants, and pedestrian gait or ambulation dynamics (English 1996). It is also generally recognized that surface slope, cross slope, and surface movement can affect slip resistance; however, the majority of published research appears to address level and stable pedestrian walking surfaces.

In this study, 36 different combinations of boot/surface/contaminant experiments were examined using a novel testing device—a dynamic shoe tester designed to simulate parameters of human ambulation on normal walking surfaces. The device is named the English XST Traction Tester (US patent 5 259 236). An analysis of variance (ANOVA) was performed on the raw sample data to detect interactions and significance levels. The test slip index readings were compared by surface contaminant, deck coating, and footwear, and any interactions were noted.

A common theme among more modern slip-resistance studies is the attempt to correlate surface roughness measurements to dynamic friction properties. While it is commonly known that floor and shoe texture properties affect slip resistance, more modern research seems to be directed at better measuring the specifics of surface roughness. Current researchers, such as Chang, Grönquvist, and Bunterngchit, seem to generally agree that a strong relationship exists between surface roughness measurements and slip resistance results. The methods for measuring surface roughness and surface characteristics can also vary widely. The hardness characteristics of walking surfaces and footwear bottoms are generally not taken into account in slip resistance research. Facility owners, such as a grocery store chain for example, may have little control over pedestrian footwear. Therefore, much US research tends to be litigation-driven and geared toward floor surfaces rather than ideal combinations of floor and footwear. With the exception of Grönqvist (Gronqvist, Roine et al. 1990), little published research has been directed toward mariner ambulation and slip resistance aboard work vessels. Because the crews on board a fishing vessel are "captive" in the sense they can't leave the work space, this situation presents an ideal environment where slip resistance research concerning deck surface-footwear combinations could be of benefit.

Materials and methods

This study was designed to explore the relationship and causal factors between several variables affecting slip resistance. It concentrated on a limited number of deck surfaces and work boots under varying environmental conditions simulated in a laboratory setting. Pedestrian gait dynamics for personnel aboard marine vessels was not a consideration in this study as no reliable reference concerning the gait dynamics of marine personnel could be found, and a test device that mimics such pedestrian gait dynamics (specific to marine deck locomotion) could not be located.

Numerous slipmeters have been designed or developed to assess floor slipperiness. These slipmeters can produce quite differing results, and some may not provide reliable data when used on contaminated surfaces (Chang 1999). Most slipmeters are designed for testing floor surfaces only and do not allow for the incorporation of footwear. A small number of machines reported in the literature allow for the use of footwear during testing. The demand for a traction testing device to rate the performance of footwear on contaminated surfaces is present but is complicated by fundamental problems that include, the lack of a recognized device and standards for its use. The slipperiness of deck coatings and footwear combinations was assessed with the English XST Traction Tester (Figure 1).

The principle of operation for slip resistance measurement by the device is calculated by measuring the tangent of the angle of incidence at which the test boot will slip when brought into contact with the test surface. The slip resistance measurement is referred to as the "slip resistance index" rather than a static or dynamic coefficient of friction. The device gives test indications in degrees of angle-of-incidence or as a slip resistance index (Figure 2).

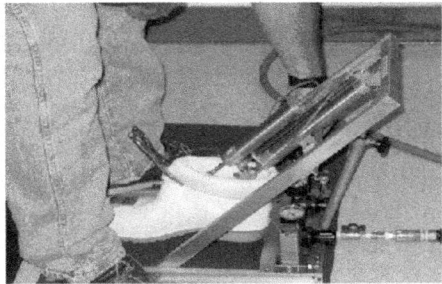

Figure 1: English XST Footwear Traction Tester and boot

Figure 2: Slip index scale

Marine deck coating samples were applied to identical sheets of mild carbon deck steel cut to 36 by 18 inches to facilitate test standardization. The four deck coatings used in the study are commercially available and are formulated and marketed for the marine industry. Three are marketed as slip-resistant coatings or additives to coatings, while one was simply marine deck paint. The coatings were chosen to represent different types of materials used to increase the roughness of marine deck surfaces. All coatings were applied according to the manufacturer's specifications.

The following summarizes the four types of deck coating selected.

1. Painted steel with no aggregate or textured material incorporated.
2. Painted steel with aluminum oxide grit mixed with paint.
3. Painted steel with rubber-crumb grit mixed with paint.
4. Steel coated with a spray-foam application of polyurethane (also commercially marketed as a pickup truck bed spray liner under the name Rhinoliner).

Work boots selected for the study are all commercially available work boots designed for industrial or marine use. Three boot styles were selected for evaluation and are generally classified as follows:

1. All-purpose steel-toed industrial work boot with oil-resistant lug sole.
2. Soft-soled PVC work boot (known locally as a white "shrimpers" boot).
3. Soft-soled, industrial polyblend work boot impregnated with strips of "renewable" aluminum oxide grit (marketed as slip-resistant boots).

Three levels of contaminants with increasing levels of viscosity were chosen for the study to test the boot/coating combinations under varying environmental conditions.

1. Dry (no contaminant applied).
2. Wet (distilled water).
3. Oil (SAE 20).

All contaminants were applied manually with spray bottles until an even film of contaminant covered the test surface. Test boot sole and heel surfaces were also evenly contaminated prior to each test run.

Test procedure

All test procedures were performed at the University of Southern California in the Department of Biokinesiology and Physical Therapy's Musculoskeletal Biomechanics Research Laboratory. The procedure required 36 different combinations to be tested. Each test combination (coating-boot-contaminant) was repeated four times with the direction of boot travel parallel to the longest side of the test surface. The tests were then repeated four times with the direction of boot travel parallel to the shortest side of the test surface. This provided a total of eight test results for each test combination for a total of 288 tests. The test results were then averaged for each test combination.

The test device incorporates a cycle timer so that the same forces and force duration are applied to each shoe tested. To initiate the testing sequence, a plunger on the test device is manually depressed. The device will then proceed through an automatic test cycle. The test cycle causes the test boot to contact the test surface in a heel-first manner. The test boot rotates about a simulated "ankle" joint until it rests flat on the test surface. A horizontal force, along with the vertical force, is then applied, causing (or not) the footwear to slide forward along the test surface. The testing cycle for each test combination was repeated until a slip occurred four times. The test device was then rotated 90°, and the test sequence was repeated until a slip condition occurred four more times in the perpendicular direction.

The test device has an adjustable frame that allows the test boot angle of incidence to be manipulated with a set screw. The frame was adjusted during the test sequence until the angle of incidence was high enough to cause a slip. This slip was quite evident as shown by the continued forward motion of the test boot in the testing device. Because of the relative crudeness of the frame mechanics, all slip index readings were rounded to the nearest 0.05. The test device measurement gauge indicated an angle of incidence or slip index. Calculating the tangent of the angle of incidence derives the slip index figure.

Statistical analyses
ANOVA was performed on the raw sample data to detect interactions and significance levels. The test slip index readings were compared by surface contaminant, deck coating, and footwear, and any interactions were noted.

Results and discussion

The test results for the surface-boot combinations under dry conditions show that all test combinations failed to produce a slip. Since most slip accidents are related to contaminating conditions, it is not surprising that the dry surfaces would not be considered "slippery." The test data do not suggest that the combinations will not cause a slip, only that the test device limits were reached prior to reaching a slip condition.

All surface-boot combinations but two failed to slip up to the limits of the test device. Not surprisingly, both slips occured on plain painted steel. Data for the wet tests also show that when a deck surface was treated with one of the slip-resistant coatings, results rose significantly.

While surface roughness measurements were not a part of this experiment, it was assumed that plain painted steel would have the lowest surface roughness figure. Crude measurements of surface aggregate peak-to-valley distances were taken using a dial micrometer and indicated that surface 2 peaks averaged 127 μ, surface 3 averaged 254 μ, and surface 4 averaged about 152 μ. Surface 1 was relatively smooth and glassy, averaging a peak-to-valley distance of 25.4 μ. Although surface 3 appears to have the highest peak-to-valley surface characteristics (approximately 254 μ), it did not perform as well as the two test surfaces (2 and 4) and had less pronounced peaks. This could be attributed to other surface roughness characteristics not measured or to some compressability factor of the rubber crumb aggregate in the surface coating. While the rubber crumb aggregate gave improved results over plain painted surfaces, the crumb material did not appear to be hard enough to sufficiently engage the soles of the test boots to prevent a slip. This type of surface aggregate is commonly used in water slide and other amusement parks where patrons spend a good deal of time barefoot. The ability for soft rubber crumb to "engage" a bare human foot may be quite different than for engaging a hard boot sole.

Most combinations of surface and boot produced a slip condition during oil contamination down to the lower detection limits of the test device. Three test combinations failed to produce a slip at the device's upper detection limit. Oily contaminant performance on surface 4 appeared to be excellent with boots 2 and 3. Two combinations produced mid-range values when subjected to oil contamination. During testing, boot 3 grit strips absorbed the oil contaminant and grit particles were easily abraded from the boot surface, so that they acted as a type of ball bearing to actually increase slips. These two factors likely contributed to poorer performance on most surfaces during the oil tests.

In these series of tests, three test combinations failed to produce a slip under all three contaminating conditions. These three superior combinations were traditional footwear and nonskid paint, grit-impregnated boots, and foamed polyurethane deck coating. Specialized boot surfaces or deck coatings appeared to offer practical alternatives to traditional deck surfaces to help provide better traction while performing work at sea. Two of the polyurethane coatings tested also offered a high degree of corrosion and chemical resistance. Nontraditional deck surfaces, both above and below the weather deck, can offer a high degree of slip resistance and greater comfort while walking or standing or when a fall does occur.

This study illustrates that the traction characteristics of marine deck walking surfaces can be evaluated and improved with footwear and surface combinations in a laboratory setting. Some of these evaluation techniques and test equipment have only become available in the last few years. Likewise, the market has seen a flood of new products claiming to improve slip resistance.

For a vessel owner or builder, more choices are now available regarding decking surfaces and crew footwear.

Footwear and surface combinations can only be compared to one another within the scope of this study, since the test device used is different than those used in other studies in the published literature. This method may prove useful to the marine industry, particularly if used in conjunction with subjective human evaluations of footwear and deck coatings. Factors that may be pertinent to material selection for marine businesses, such as cost and durability, were not evaluated or considered in this particular study; however, all the materials used are commercially marketed for use in marine industries, are relatively cost-effective, and are durable in industrial marine settings.

While general knowledge in the marine industry holds that grit coatings, such as surface 2, can provide a more slip-resistant deck surface, these coatings can abraid equipment, such as fishing nets, and catch if sorted from the deck. These surfaces can also be hard and abrasive to personnel if someoe does fall. Specialized surface coatings, such as 4, appear to offer a practical alternative for marine decks when used with appropriate footwear and would likely be less abrasive to equipment or personnel.

The test conditions in this study were designed to simulate actual environmental conditions that may be found in a marine environment. Several useful test devices are available that could be used under actual field conditions in future studies. The results of this study offer information useful in slip resistance research, but cannot explain all possible interactions. A number of questions remain that offer many possibilities for expanded research. Recommendations for further study include refinements to the line of investigation presented in this study and development of new analysis techniques for footwear and surface combinations on marine work surfaces. Research should also include ambulation studies of work aboard marine vessels, development of improved slip-resistance testing devices, expansion of the types of surfaces and footwear tested, and further exploration of roughness and compressability factors on nontraditional deck surfaces. A comprehensive investigation into the incidence and causes of slips and falls aboard marine vessels can also help guide future maritime tribometric investigations.

Acknowledgments

The laboratory tests for this study were conducted at the University of Southern California, Musculoskeletal Biomechanics Research Laboratory. Special thanks are given for use of the laboratory and equipment during the course of this research. The East Carolina University School of Allied Health, Department of Biostatistics, assisted with the statistical analysis. Although the study itself was self-funded, the Coast Guard provided and funded numerous opportunities for research networking as well as the presentation of study results.

References

Bunterngchit Y (2003). The influence of floor surface roughness on the slip resistance coefficient. Available at http://library.kmitnb.ac.th/article/atc34/atc00136.html.

Chang WR (1998). The effect of surface roughness on dynamic friction between neolite and quarry tile. *Safety* 29(2): 89-105.

Chang WR (1999). The effect of surface roughness on the measurement of slip resistance. *International Journal of Industrial Ergonomics* 24: 29-313.

English, W (2000). Footwear safety and traction in the workplace. *Professional Safety* April: 23-26.

English W (1996). Pedestrian slip resistance: How to measure it and how to improve it. Alva, FL: William English, Inc.

Gronqvist R, Roine J et al. (1990). Slip resistance and surface roughness of deck and other underfoot surfaces in ships. *Journal of Occupational Accidents* 13: 291-302.

Lincoln JM and Conway GA (1997). Commercial fishing fatalities in Alaska: Risk factors and prevention strategies. Cincinnati, OH: National Institute for Occupational Safety and Health. Current Intelligence Bulletin 58. DHHS (NIOSH) Pub. No. 97-163.

National Institute for Occupational Safety and Health (1994). Preventing drownings of commercial fishermen. DHHS (NIOSH) Pub. No. 94-107.

US Coast Guard, Fishing Vessel Casualty Task Force (1999). *Living to Fish, Dying to Fish*.

A VESSEL SAFETY AND SEAFOOD PROCESSOR ORIENTATION PROGRAM FOR WESTERN ALAS-KANS

James Herbert
Alaska Vocational Technical Center
Seward, Alaska, USA

Jerry Ivanoff
Norton Sound Economic Development Corporation
Unalakleet, Alaska, USA

Abstract

Federal legislation established the Community Development Quota (CDQ) program to allow Bering Sea communities to share in the commercial harvest of many species. The Norton Sound Economic Development Corp., representing 15 villages, formed a partnership with the Alaska Vocational Technical Center to train area residents for processing jobs on factory trawlers that harvest pollock and other species of fish. This paper describes the CDQ program, the marine safety and seafood processing training program, and the benefits it has provided to Norton Sound residents.

Introduction

The Bering Sea is one of the most productive parts of the world's oceans. It is estimated that 20% of the fish directly consumed by people on the planet comes from its bounty. This amounts to some 1.5 million tons of fish per year (Witherell 2000). Unfortunately, even after the transitions from foreign harvest, to joint ventures, to the largely American harvest of the resource late in the 20th century, people in the communities of western Alaska adjoining the Bering Sea did not see much economic benefit. With many of the corporate interests and larger vessels based outside Alaska, there was little incentive to include economically depressed regions of western Alaska in the decisions concerning resource management of the Bering Sea.

Community Development Quota program

After hard work by Alaska political leaders and strong support by the Alaska Congressional delegation, a plan to share some of the bounty of the Bering Sea with the region's residents evolved. The Community Development Quota (CDQ) program was begun in December 1992 (State of Alaska 2003) and formalized by the US Congress under the Magnuson-Stevens Fishery Conservation Act (1992). In the initial CDQ program 7.5% (approximately 100,000 metric tons) of the annual harvest of pollock in the Bering Sea was allowed to be harvested by entities representing the people living in communities bordering on the Bering Sea. The program stipulated that the communities that lay within 50 nautical miles of the Bering Sea form associations to arrange for the harvest of the pollock resource. They were to use the proceeds of the fishery for the economic benefit of the people in the region.

The original 56 communities formed six associations. Based on the number of persons represented by the association and analysis of business plans submitted to the State of Alaska Department of Community and Regional Affairs, a share of the CDQ allocation was granted to each association. Careful oversight by the State of Alaska and the U.S. Department of Commerce mandates accountability.

The initial program demonstrated success, and it has evolved over the years. There are now 65 villages participating. The CDQ share of the annual pollock harvest has increased to 10%. There is now also a CDQ allocation of other Bering Sea species, including king crab, tanner crab, Pacific cod, halibut, and sablefish. While the program is not without its critics, there is no doubt that appreciable economic benefits have come to the region. In 2002, 1,900 CDQ jobs brought $12 million in wages to western Alaska (State of Alaska 2002).

The 15 villages in the Norton Sound region of Alaska formed the Norton Sound Economic Development Corporation (NSEDC) to pursue economic development based on their allocation of fishery quota. It established a partnership with the Glacier Fish Company of Seattle to catch much of the resource quota. The relationship has proven successful to the point that NSEDC now owns 50% of the Glacier Fish Company.

Since a significant objective of the CDQ program is to provide gainful employment to western Alaska residents, NSEDC looked for job opportunities aboard factory trawlers. While many residents of the region rely heavily on subsistence resources and some are engaged in near-shore commercial fisheries on small vessels, few individuals had experience or training for the distant-water fisheries practiced by the large factory trawlers, which generally had their home port in Seattle.

Vessel safety and seafood processor orientation program

Working with the management of Glacier Fish Company, the NSEDC Board of Directors determined careful screening and training of candidates would be a precondition for employment aboard the factory trawlers. Three times each year the training coordinator of NSEDC recruits potential workers. Each village has a local representative who can talk to people and provide information. Applications are forwarded to the training coordinator who makes the selection of who will participate in training.

The chosen individuals are brought to Nome, Alaska, as a central city in the region. To reinforce NSEDC and Glacier Fish Company's zero tolerance of illicit drug use, everyone is given a pre-employment drug screening. The Coast Guard mandates zero tolerance for illegal drug use aboard all US commercial vessels, and so the rules are set for workers from the start. Since the nature of the work is strenuous, individuals are also given fitness-for-duty physical examinations.

Many candidates have little experience outside the region and the whole process is meant to prepare them for what lies ahead. Accordingly they travel by airplane and then bus from Nome, to Anchorage, to Seward to begin the training process.

The Alaska Vocational Technical Center (AVTEC) is located in Seward and is part of the State of Alaska Department of Labor and Work Force Development. NSEDC has formed a training partnership with AVTEC because of a number of factors. The Maritime Department has more than 10 years of experience orienting CDQ and specifically NSEDC trainees for work on factory processors. Trainees must adapt to dormitory life and cafeteria food in a new town away from friends and family. This is similar to changes they will experience with shipboard life. Strict rules about drug and alcohol use are enforced as they would be on board the Glacier ships. Students must adhere to a regular class schedule. The training is varied but focuses on four general subject areas: Safety and Survival at Sea, First Aid and CPR, Commercial

Fishing Methods, and Fish Processing Technology. Competencies are clearly defined at the start of classes, and employment is preconditioned on successful completion of the 2-week class. Hands-on skills are emphasized as much as possible.

It is hoped that those who successfully complete the training process have the self-discipline to adapt to new circumstances and not be distracted by alcohol. They should have a solid knowledge of personal survival equipment, first aid, industrial safety, commercial fishing methods, and processing methods. Individuals should also have a sense of how well they are suited to the rigors of shipboard life and the duties of a seafood processor. While people are strongly encouraged to complete the training program, they are under no obligation to go to sea. NSEDC feels many of the skills taught in the orientation program, such as cold water survival and first aid, will have application to the lives of people whether at home in the village or on a factory trawler.

At the end of the 2-week orientation program, a human resources manager from Glacier Fish Company interviews those persons interested in at-sea employment. They fill out applications and federally required paperwork. Individuals are given an approximate schedule of when there might be job openings for new employees, as well as company policies on travel, pay, and advancement. At this point, motivated individuals should be prepared for the ship-board workplace and will bring a realistic but enthusiastic attitude to the job.

The class schedule followed while the trainees are in Seward is shown on the next two pages. A typical school day runs from 0830 to 1700. The various components of the class are taught in blocks with the intent that immersion is the best way to enhance learning.

The first day begins with an orientation to the facilities and schedule. School rules and competencies required for successful course completion are presented. Since many individuals are unfamiliar with nautical vocabulary and parts of a ship, lectures and exercises deliver this material early in the course. Relevant videos are often used throughout the course to emphasize key points.

The module of the course dealing with Safety Equipment and Survival illustrates the relationship between lectures and hands-on skills. As shown in the schedule, a variety of topics is covered. Reference material is provided by is-

ALASKA VOCATIONAL TECHNICAL CENTER CLASS SCHEDULE

Seafood Processing and Marine Safety Orientation Class
WEEK ONE

Time	Monday	Tuesday	Wednesday	Thursday	Friday
SESSION 1	Introductions, orientation to facilities + staff, schedule, competencies, expectations	Preparation for an Emergency Seven Steps to Survival	Life Raft Equipment and Launch	Fire Fighting	Personal equipment and supplies for a two month voyage
SESSION 2	Overview of the history and intent of the CDQ program and NSEDC opportunities	Hypothermia Coldwater Near Drowning	Signals EPIRB	Fire Fighting	Ergonomics and proper lifting techniques
	LUNCH	LUNCH	LUNCH	LUNCH	LUNCH
SESSION 3	Bering Sea Nautical vocabulary	Personal Floatation Devices Immersion Suits	Radio communications MAYDAY	Fire Fighting with hand held extinguisher and flare exercise	Industrial Safety
SESSION 4	Layout of Factory Trawlers Parts of a ship On board life	Immersion suit Practice Life Raft	Helicopter Rescue	Station Bill and On Board Drills Safety Exam	Industrial Safety
HOME WORK	Reading	Pool Session	Reading Review for Exam	Prepare list of personal equipment for a two month voyage	First Aid Reading

ALASKA VOCATIONAL TECHNICAL CENTER
CLASS SCHEDULE

Seafood Processing and Marine Safety Orientation Class
WEEK TWO

Time	Monday	Tuesday	Wednesday	Thursday	Friday
SESSION 1	First Aid	First Aid	First Aid Exam Commercial Fish Species	Processing Operations	Product Quality Control
SESSION 2	First Aid	First Aid	Long lining	Processing Operations	Surimi and Meal Production
	LUNCH	LUNCH	LUNCH	LUNCH	LUNCH
SESSION 3	First Aid Skills	First Aid CPR	Demersal Trawling	Product Quality Control	Processing Exam Course Evaluations
SESSION 4	First Aid	First Aid	Mid-Water Trawling	Field Trip	Presentation of Certificates Job Interviews with Glacier Fish Company
HOME WORK	Reading	Review for First Aid Exam	Reading	Reading and Review for Exam	Finished

suing each student a copy of the book *Beating the Odds in Northern Waters* (Clark-Jensen and Dzugan 2002). Nightly reading is assigned. In one section of this module, lectures and discussion cover preparation for an emergency, hypothermia, cold water near-drowning, and person overboard. Emphasis is placed on the use of immersion suits as a key method of enhanced survival in cold Alaska waters. After discussion and demonstrations, each individual is issued an appropriately sized immersion suit. They are required to inspect the suit for material damage, lubricate the zipper, and practice donning it. For nearly all trainees, this is their first exposure to this critical piece of gear, and some need more practice and guidance than others. As illustrated by the competency checklist, the students must be able to don the suit in the classroom in less than 2 minutes, although with practice, nearly everyone is able to do it in less than a minute in competitive time trials. While not timed, everyone also needs to don a suit in the dark to demonstrate familiarity with the gear.

In the controlled environment of the local high school swimming pool, the trainees are able to gain confidence in the buoyancy of immersion suits and various personal flotation devices (PFDs). In-water skills minimally include swimming a distance with an immersion suit, properly entering the water from a height of 1 meter while wearing a suit, climbing into an inflatable life raft from the water, righting a capsized life raft, and comparing the comfort and effectiveness of various PFDs in the water.

While some individuals are excellent swimmers, others have had negative in-water experiences or are nonswimmers. The goal is to have all trainees gain confidence in the equipment that could save their lives in an emergency. It is extremely gratifying for the AVTEC maritime instructors to see a nonswimmer overcome his or her fear and trepidation of being in the water and use the equipment properly. Because most ships conduct drills monthly (US Code of Federal Regulations 2002), there is regular practice in donning immersion suits, but this pool exercise is one of the few times individuals get to use suits and rafts in the water. Students must dry and maintain all the equipment they use in the pool. This time-consuming task emphasizes the importance of proper care for safety equipment.

First aid and cardio-pulmonary resuscitation (CPR) are skills that may be useful to people in all phases of their lives. It is an aspect of the safety training that individuals can make use of on the vessels and at home in the village. This module is based on a Coast Guard-approved first-aid and CPR course.

Students who successfully pass a written test and demonstrate the skills listed below are issued a card valid for 2 years. The intent is to help individuals render immediate assistance until a person of greater medical training and ability arrives on the scene. The students can then be of assistance to that person. Students are more enthusiastic about training when they are actively engaged in performing skills on manikins or each other.

The material presented in the Safety and Survival and First Aid modules of the course is similar to those modules in the International Maritime Organization's (IMO) Basic Safety Training course (International Maritime

---

NORTON SOUND ECONOMIC DEVELOPMENT CORPORATION

SEAFOOD PROCESSING AND SAFETY ORIENTATION CLASS

REQUIRED COMPETENCIES

XX PUT ON A SURVIVAL SUIT IN LESS THAN TWO MINUTES

XX PUT ON A SURVIVAL SUIT IN THE DARK

XX SWIM 50 YARDS IN A SURVIVAL SUIT

XX RIGHT A LIFERAFT WITH A SURVIVAL SUIT

XX ENTER A LIFERAFT WITH A SURVIVAL SUIT

XX CLEAN, DRY, AND MAINTAIN A SURVIVAL SUIT

XX DEPLOY AT LEAST ONE FLARE

XX MAYDAY (WRITE OUT OR STATE FROM MEMORY)

XX EXTINGUISH A DIESEL FIRE WITH A HAND-HELD EXTINGUISHER

XX SHOW LIST OF PERSONAL EQUIPMENT & SUPPLIES FOR A 2-MONTH VOYAGE

XX WATCH VIDEOS IN FISHERIES SURVIVAL SERIES

XX WATCH VIDEOS IN NPFVOA SAFETY AND SURVIVAL SERIES

XX WATCH VIDEOS IN AFTA INDUSTRIAL SAFETY SERIES

XX ATTEND U.S.C.G. APPROVED FIRST AID COURSE (CARD ISSUED: YES / NO)

XX ATTEND C.P.R. COURSE (CARD ISSUED: YES / NO)

XX PASS A WRITTEN TEST ON SEAFOOD PROCESSING AND TRAWLER ORIENTATION WITH A MINIMUM SCORE OF 70%

XX PASS A WRITTEN TEST ON MARINE SAFETY WITH A MINIMUM SCORE OF 70%

XX HAVE NO MORE THAN FOUR SESSIONS OF UNEXCUSED CLASS TIME

XX OBSERVE ALL SAFETY RULES

STUDENT	INSTRUCTOR	DATE

Organization 1997). This program provides more time in lectures and skills practice than the minimums suggested by the IMO. Because many individuals have had little or no marine safety training, it is felt this provides a solid foundation for course objectives and cultivates a positive attitude toward safety.

Students are given an overview of relevant commercial fishing methods for species they are likely to encounter. Basic fish anatomy and identification are presented as well. As seafood processors, the main focus of their work

Elementary First Aid Course
Required Competencies
AVTEC - Seward Alaska

Student's Name: _____

Practical Exercises

____ Adult One-Person CPR

____ Adult Rescue Breathing

____ Adult Concious Choking

____ Adult Unconcious Choking

____ External Bleeding Control

____ Shock Position

____ Recovery Position

____ Clothes Drag

____ Lifting Technique

____ Log-Roll

Written Examination
____ Score a minimum of 70% on a written exam.

Classroom
____ Have no more than one (1) hour of unexcused class time.

____ Observe all safety rules.

Assessor   Initials

Date completed: _____   _____   _____

will be on producing quality food products at sea. Treating all fish caught as the food someone will be eating sets the stage for the importance of quality. Basic bacteriology and the theory of autolysis are described, as is the critical nature of temperature in the spoilage process.

Trainees are lectured on the various fish processing machines with which they will be working and the different jobs on the processing line. The season of the year and market forces will dictate the product being emphasized. For example, pollock roe production is the focus in January with varying percentages of fillets and surimi throughout the rest of the year. Meal production is also described.

NSEDC has been committed to this safety training and processor orientation program for more than 10 years. Six-hundred individuals have participated in this program over the years. With successful course completion, the trainees have the option for at-sea employment on Glacier Seafood ships. Some chose not to go to sea, while others are enthusiastic about the opportunity. Some make a few trips and lose interest, but some Norton Sound residents mesh well with the work and integrate regular trips into their lifestyle. The cash flow to the region from seafood processor wages has been significant. The most recent figures indicate more than $8 million in direct worker wages have flowed into the region since NSEDC began preparing people through this training program. Because of the low availability of jobs in most area villages, seafood processing jobs on Glacier ships represent an important economic option to motivated and trained individuals.

Conclusions

The NSEDC Seafood Processing and Safety Orientation program provides an opportunity to screen individuals interested in employment opportunities aboard the Glacier Fish Company factory trawlers. Often individuals have had little experience outside their village and no marine safety training. The skills-based course and overview of what they can expect in terms of daily life at sea readies individuals for jobs at sea. Many of the skills learned and demonstrated in the program, such as first aid and water safety, prove useful in their day-to-day lives when individuals return to their communities. This program has qualified individuals for employment that has yielded important economic benefits to the Norton Sound region. With careful management of the fishery resource in the Bering Sea and continued dedication to the training and employment goals of the Community Development Quota program, Norton Sound Economic Development Corporation will provide economic benefits to the people of the region.

References

Clark-Jensen S and Dzugan J, eds. (2002). *Beating the odds in northern waters. A guide to fishing safety.* University of Alaska Sea Grant.

International Maritime Organization (1997). Model Course 1.19: Personal survival. London: 40.

Norton Sound Economic Development Corp. Education employment and training program. Available at http://www.nsedc.com/eet.html

State of Alaska (2002). CDQ Employment 1993-2002. Available at http://www.dced.state.ak.us/cbd/CDQ/pub/cdqstats_employment.pdf

State of Alaska (2003). Community Development Quota Program. Available at http://www.dced.state.ak.us/cbd/CDQ/cdq.htm

US Code of Federal Regulations (2002). Chapter 46: Shipping, Part 28: Requirements for Commercial Fishing Industry Vessel. Revised October 1, 2002.

US Congress (1992). Magnuson-Stevens Fishery Conservation Act of 1992.

Session Five

SESSION FIVE:
REGIONAL APPROACHES TO FISHING SAFETY

Fishermen in artisanal craft in Bay of Bengal (Photo courtesy of Dr. Yugraj Yadava)

IMPROVEMENT IN FISHER HEALTH AND SAFETY THROUGH POVERTY ALLEVIATION: A CASE STUDY ON THE INDUS BASIN FISHING COMMUNITIES IN PAKISTAN

Muhammad Naeem Khan, PhD
Mohammad Yamin Janjua
Department of Fisheries and Aquaculture
Faculty of Fisheries and Wildlife
University of Veterinary and Animal Sciences
Lahore, Pakistan

Abstract

The socio-economic conditions of traditional fisher communities in Pakistan are studied with special reference to the need for health and safety education for artisan fisheries in the Indus basin. The study indicates that poverty alleviation not only contributes to the welfare and prosperity of these fisher communities, but also promotes safety and occupational health by enhancing awareness. Supplementation of traditional fishing with aquaculture to diversify resource competition is suggested as a means of improving the socio-economic conditions in fisher communities. Health and safety issues are discussed as related to both small-scale fisheries and the aquaculture industry, including a variety of common hazards, from boat integrity to electrical safety, and from drug and chemical handling to WHIMIS training. Improvement and/or gradual replacement of primitive and unsafe fishing gear and technologies, coupled with education and training are suggested to promote fisher health and safety. Community participation and the role of nongovernmental organizations (NGOs) in promoting health and safety awareness are discussed with special reference to local cultural traditions of Southeast Asia. The paper will review the environmental disaster in the wake of the oil spill caused by the ship *Tasman Spirit*, which ran aground in August 2003 at the port of Karachi, and its impact on fisher health and the economy. This oil spill has devastated the coastal fisher communities who earn their livelihood from fishing in the Arabian Sea coast.

Introduction

Pakistan

Pakistan, a maritime nation, is endowed with rich fishery resources and potential. Pakistan is located along the northern shores of the Arabian Sea and has a coastline of about 1,100 km, with a broad continental shelf and an Exclusive Economic Zone (EEZ) extending up to 200 nautical miles from the coast. About 16,000 fishing boats operate in coastal areas in shallow coastal waters and offshore areas. These fishing boats undertake trips from a few hours to about 25 days, depending on the type of fishing. Total production from inland and marine waters is 590,000 metric tons.

Fishing plays an important role in the national economy of Pakistan. It provides employment to 336,000 fishers directly. In addition, another 400,000 people are employed in ancillary industries. It is also a major source of export income. In 2002-2003, fish and fishery products valued at US $130 million were exported from Pakistan.

Marine resources define the life and livelihood of more than two million people in the coastal provinces of Sindh and Baluchistan, spread across 1,100 km between Sir Creek, bordering India in the east, and Jiwani, adjacent to Iran in the west. The largest concentration of marine fish harvesters and workers is in Karachi Division. The remaining Sindhi population of marine fishers is located largely in Thatta District. The Baloch fishers mostly reside in Gwader, with some found in Lasbela District. A large population of migratory inland fishers live in the Indus River basin, mostly near dams and barrages.

Fisher health and safety

Fishing has been regarded as one of the most challenging and dangerous occupations in the world. The International Labour Organization (ILO) and Food and Agriculture Organization (FAO) estimate that 7% of all worker fatalities occur in the fishing industry, despite the fact that the industry accounts for less than 1% of the worldwide workforce (Lincoln, Hudson et al. 2002). The ILO has estimated that 24,000 fatalities and as many as 24 million nonfatal injuries occur worldwide each year. The fatality rate of the world's fishermen is estimated to be 80 per 100,000 workers per year (ILO estimate). In the case of commercial fishing vessels, safety is a complex interaction involving humans (skipper, crew member, owner), machines (vessel, equipment), and environment (weather, management scheme) (Lincoln,

Hudson, and Conway 2002). The building blocks for artisan fisheries may be the same, but they have not been defined as clearly. Safety problems occur when a single element–human, machine, or environment–malfunctions. Human factors can include fatigue, inexperience, or nonuse of safety equipment. Machine factors can involve older vessels, such as in the case of the *Tasman Spirit*, which spilled oil off the coast of Karachi in August 2003, as well as inadequate safety guards for heavy machinery used in many fishing operations. Environmental factors can include harsh weather and slippery, unstable work surfaces (Lincoln, Hudson, and Conway 2002).

While there is no universal solution for fisher health and safety, strategies to prevent fishermen from being injured or killed on the job need to be explored through efforts such as improving vessel stability and hull integrity, making safety equipment such as survival suits and life rafts more widely available, implementing safer management regimes, understanding and heeding weather information, and averting falls overboard (Lincoln, Hudson et al. 2002).

The FAO reported increased fatalities in artisan fisheries in developing countries as a consequence of changes in the basic nature of fishing operations: overexploitation of coastal resources; advances in vessel and fishing technologies, including motorization and new types of fishing gear; lack of training, experience, and skills; commercial pressure; etc. Where inshore resources have been overexploited, fishers are often opting to work farther away from shore, sometimes for extended periods of time, in fishing vessels designed for single-day, inshore fishing operations (Food and Agriculture Organization 2000). Small-scale fisheries are particularly vulnerable to worker injuries (Ben Yami 2002).

While the general principles and strategies for improvement in health and safety of these artisan fishers are the same as those of commercial fishers, the actual requirements for artisan fisheries can be modified according to the fishing culture and traditions in the area.

Safety problems of small-scale and artisan fisheries have received low priority, even in many industrial nations, and have been all but neglected in most of the others (Ben Yami 2002). The number of casualties is high in Pakistan and other developing countries where small-scale fish workers operate under conditions in which their vessels, safety and communication equipment, first aid, search and rescue, and early warning services are either inadequate or totally absent (Ben Yami 2002).

The consequences of loss of life fall heavily on dependants. In developing countries, the consequences can be devastating: widows have a low social standing; no welfare exists to support bereaved families; and, lacking alternate sources of income, widows and children may face destitution (Food and Agriculture Organization 2000). The present study shows that supplementing traditional fisheries with small-scale, land-based pond aquaculture provides alternate sources of income to poor fishers and bereaved families whose wage-earner perished while fishing. Small-scale integrated pond aquaculture not only diversifies the resource competition, it also helps in improving health and safety, through poverty alleviation, in these fishing communities. Experience has shown that instead of focusing on the individual fisher, the fisher community as a whole should be targeted for health and safety education. The village councils, fisher cooperative societies, and other NGOs can help governments disseminate health and safety education in these poor and largely illiterate communities. Reductions in casualties can be achieved through the concerted actions of fishing communities and organizations, national and subnational authorities, international organizations, and voluntary bodies (Ben Yami 2000).

Fishermen community safety: A Pakistan perspective

Risk to fishermen's communities

Residents of fishing communities are at direct risk of accidents and diseases because of direct contact with water, harsh environmental conditions, poor working conditions, polluted waters, and fish, which spoil easily and can carry disease. Fishermen, especially inland fishermen, live in makeshift villages on banks of rivers and reservoirs where sanitary conditions are very poor. They have no access to hygienic drinking water and are bound to drink water from reservoirs, dams, rivers, and streams, which are becoming more and more polluted by growing industrialization and human population pressures. There are no latrine facilities in these villages. Because of the lack of availability of proper medical facilities in fishing communities, mother and child mortality rates are quite high.

Poverty alleviation projects: Participation approach

Community development in fishing areas should be taken on from a grassroots, bottom-up approach, as opposed to the conventional approaches of development planning. Such an approach would entail, first, the direct involvement of small-scale fishermen in problem-solving through dialogue and, second, the participation of small-scale fishermen themselves in project formulation and implementation. Placing small-scale fishermen at the centre of the policy will not only ensure participation, but will also permit the socio-economic and cultural needs of the communities to be taken into account and their own priorities respected.

Development of fisher communities also depends on availability of resources. Projects should be all-encompassing, involving and integrating agriculture and other land-based projects.

The basic purpose of a poverty alleviation programme should be to improve income opportunities in fishing communities, both within and outside the fisheries sector, and improve the standard of living. To succeed, the programme requires strong and independent fisher organizations and cooperatives that can function as pressure groups for community development. The government has attempted to initiate changes at the level of fishing cooperatives to enable them to participate in the new development programmes. Fisher cooperatives should be used as an instrument to improve fisher welfare. There should be small and medium enterprises (SME) banks to provide easier access to credit and initiate schemes to improve the common welfare of these target communities.

The scope of such programmes should include developing fisheries resources in accordance with sound fisheries management practice; generating employment opportunities in the fisheries sector by expanding and modernizing fish production and related secondary industries; and supervising, promoting, and undertaking the economic and social development of fishermen associations.

Fishermen in inland waters are primarily hired by river and reservoir fisheries contractors on a percentage basis. The fishermen cooperatives on their own can contract with fishers, after securing loans and credit from SME banks, which increases income-generating possibilities by eliminating middlemen.

Improvement of health and safety through poverty alleviation projects

The projects should include improving the availability of safe drinking water by providing hand pumps, improving sanitary conditions by building public latrines and appropriate locations to dump fisheries wastes, promoting health facilities through visits by mobile health workers, establishing health centres, and training and educating fishers in basic health.

Community participation

A high level of participation by fisher folk is essential for promoting health and safety in fisher communities. Fishers from poor communities with low literacy rates and socio-economic conditions fish Pakistan waters. The working conditions for these people are appalling, as they have little or no access to basic civic amenities. Generally, these communities grow in isolated and neglected settlements, neighbourhoods, or villages near the coastal towns. While land-based agriculture is the profession of the majority of Pakistanis (74%), fishing is generally seen as a dirty and polluting activity.

Poor fishing communities are seldom in a position to win their own means of production, and thus they have to work on a share basis for the owners of larger boats and nets. The women often make the nets in the fishing communities on behalf of the boat owners, who can procure or invest in the cottage industry of twine making. The fishers, being largely illiterate, lack modern education and only possess traditional knowledge and skills, hindering their emancipation.

Most professional fishers own only their household plots, which are small. A few own additional cultivable land, but many are absolutely landless. It would appear that the fishing communities are very homogenous. This, however, is not the case. They all share a low-status position, but some of the fishers do own boats (although mostly smaller, local ones), nets, and land. Some are more influential with the people in the larger community who control resources, such as markets, and thus gain economic and political influence and benefits.

Assets in the fishing sector are, to a large extent, owned by the middlemen in the fish trade, "aratdars," landowners (in the more rural areas), and businessmen. With the increase in profits following the increase in mechanization and better marketing, investments in the marine fishing sector seem to be by district. This has led to an increase in the incomes of the professional fishers who are experienced in operating the larger, mechanized boats. In parts of these districts, it is said, the daily wage of a professional fisher is about 100 rupees; as a comparison, daily wages for agricultural wage labour is 30-60 rupees in the same area. But the changes have happened fast during the last few years, and too little is known about their effects and impacts.

Most of the marine fishing communities and fishing households still have to be regarded as very poor and vulnerable. The communities are not integrated in the development activities benefiting other communities, and the households very often have to operate using survival strategies.

Fishing is a seasonal activity, and this means that the majority of poor fishing families are seasonally and in times of emergency are severely short of money, to the extent that most will have to borrow money to eat. This, in turn, makes them dependent on some source of quickly disbursed credit, usually from one of the employers or traders in the fishing sector. The dependency of working fishers is reinforced, and, in terms of working relations and trade, is of advantage to employers and intermediaries. Any strategy to improve the terms of trade for poor fishermen, or increase their incomes, must take such factors into account.

Women in fisher communities

Fisher communities are somewhat liberal compared to other communities and tribes in Pakistan and could be considered to be female dominated. The main city and port of Pakistan, Karachi, was initially inhabited by fishers and was named after an elderly fisherwoman, Mai (lady) Kolachi, who sold fish oil to neighbouring countries. Interestingly, children in fisher communities are called by their mother's name instead of their father's. Neighbourhoods, too, are named after mothers. The mother is in charge of all the family matters, including matchmaking and wedding arrangements, going to the markets, and looking after the cattle when the men are away at sea during the fishing seasons. They make handicrafts for home use and sale as an additional income source. Women and their families also accompany the men on sea-fishing expeditions. They participate in preparing fish nets and shelling shrimp. In Sindh Province, on Kinjher Lake, and in the Indus Delta area, fishermen live in traditionally built houseboats. All activities, ceremonies, and celebrations are confined to water.

Role of NGOs in fishermen's community development

The increasing concern with poverty alleviation for poor fishers, with popular participation in development activities, has brought attention to the role played by many NGOs. There are only a few NGOs working for the betterment of fishing communities. The biggest one is the Pakistan Fishermen Co-operative Society, which is actually under government control. The most active and popular one is the Pakistan Fisherfolk Forum. There are few others that confine their activities to specific geographic areas, including Banh Beli (working in the Indus Delta area), Sawani Sanjh (working at the River Ravi), the Sustainable Aquaculture Development Foundation (SADAF) (working in the Punjab area), the Human Rights Commission, and the Peoples Right Movement (Karachi area).

The government has realized that poverty alleviation and development in fisher communities cannot be successfully achieved without the popular involvement of communities and the NGOs. The development experience of the last 55 years has proven that NGOs have been more successful in reaching the target groups than more formal government development projects. The NGOs have been able to be more innovative and flexible in working within Pakistan's feudal agrarian society. Many of them have also been more successful in identifying the needs of the target groups and in finding ways to improve socio-economic conditions. The smaller area of focus and lesser bureaucracy helps NGOs create programmes with fewer formalities than governmental programs. It is suggested that international donors should strengthen the NGOs and university faculties to train and promote fisher health and safety in poor fishing communities in Pakistan.

Improving fishing conditions
Poverty among fishers could be alleviated by introducing aquaculture techniques in coastal areas, as well as along inland waters. This includes shrimp and finfish farming in coastal areas and in brackish water; cage and pen culture in lakes, reservoirs, and dams; and fish farming in river plains. Efforts should be made to change traditional fishing gear, crafts, and techniques to modern ones. During the last 15 years, aquaculture in Pakistan has increased at a rate of 8% per annum, especially near the Indus River basin.

Building large, small, and minidams is among the government's top priorities. These dams could provide water for irrigation, on one hand, and also become a source of income for fishers, leading to improved socio-economic conditions in the fisher community along the Indus Basin. Special attention is being given to study the projects' effects on fisheries and the environment. Regular stocking is being done in existing reservoirs and barrages by government agencies and private investors, and fish production from these reservoirs has increased significantly in recent years. At this same time, however, coastal fish farming is a neglected area and requires immediate attention to train fisher communities in aquaculture and new fishing methods.

Boat building and mechanization

In the aftermath of the 2003 monsoon cyclone that destroyed many fishing boats and caused scores of fatalities on the open sea, Pakistani fishers looked toward international assistance to address the grave issues of health and safety. An international project to build and distribute improved motorized fishing boats is part of an effort to encourage commercial artisan fishing, improve boat and motor technology, and improve the income of poor fishers, who are routinely hit by cyclones. The government has provided facilities for training in boat building at Karachi, and similar facilities are planned in Baluchistan Province. In addition, however, improvements in supplies, services, and marketing of the catch from fishers need immediate attention.

Case studies

Plight of shrimp peelers

The peeling industry is predominantly made up of women. The shrimp perish fast when warmed, so peeling has to be done prior to noon time—working hours are mostly from 4 a.m. to midday. There are no set working hours for the workers. Peeling is performed primarily with bare hands. Rubber boots, gloves, or any other protective clothing are unheard of. For this reason many of the workers' hands are covered with cuts, but for fear of losing their jobs, the workers hide these wounds. The working environment is damp and dingy. Most common health problems are back pain, swollen hands and feet, infections, and allergies. Sitting in an unchanging position for long hours causes many of these workers to develop spinal column trouble. This problem can prove very serious for pregnant women.

Oil spill: An environmental disaster for fisher communities

The running aground and final breaking up of the 24-year-old oil tanker *Tasman Spirit* in August 2003 caused a major environmental disaster in the heart of the coastal fishing communities of Pakistan. The oil was imported by a government-owned refinery in an old, single-hulled ship chartered by the Pakistan Shipping Company. The oil spillage along the Karachi coast continues to threaten Pakistan's US $130 million annual seafood export trade. The Marine Fisheries Department, Karachi Port Trust (KPT), Karachi Fish Harbour Authority, and other agencies concerned failed to control the oil spill. The fishers were directly exposed to the dangerous crude petroleum vapours, and they are feeling respiratory, eye, and skin problems.

Presently, the Karachi Fish Harbour caters to the needs of around 2,300 boats with an annual catch of around 650,000 tons of fish and shrimp.

Through export of seafood, most of which goes to European countries, the harbour earns around 150 million dollars annually. The first victim was the Harbour. The spill entered and filled the entire harbour area, creating pollution and environmental hazards for the fishers. Located at the far end of the Karachi port toward the West Wharf, the harbour had started giving off the foul smell of the oil. On the other side, concerned authorities had issued directives to the fishers not to light any fires, something they normally do to cook their food.

Since the fish and shrimp breeding season was still on, most of the fingerlings, which normally stay near the coastline, may have been killed. The negative impact of the disaster may not be immediately visible on marine life, but next season's harvesting of fish and shrimps could disclose the true story. The immediate effect would be on crab, which once lived near the coast.

Authorities used chemicals to contain the spill. The chemicals would settle the oil on the sea bed, directly affecting marine life, particularly shrimp, underneath the water. Although the Karachi Fish Harbour was already polluted, the oil layer over the sea water was more hazardous and damaging to the ecology and marine life than previous levels of pollution.

Foreign buyers are losing confidence in Pakistani seafood as more than 20,000 tons of crude oil has been spilled into the sea, damaging the mangrove forest, coral reefs, and breeding grounds of thousands of tropical and other fish species. The coastal fisher communities and artisan fishers are devastated by this environmental catastrophe. It has resulted in health problems and disease, a drastic decline in fish catches, and deterioration in socio-economic conditions of this vulnerable community. The fisher communities along the coastline where the oil tanker *Tasman Spirit* broke apart have complained of severe irritation to their eyes, respiratory problems and breathlessness, headache, drowsiness, choking, and throat irritation.

The extensive damage done to marine life and ecology needs to be evaluated, penalties need to be imposed on the responsible parties, and insurance damages for the fishers and environment be realized.

Dispersants, booms, and skimmers

Pakistani authorities were ill prepared for a possible salvage operation of the grounded oil tanker. The tanker grounded on July 27 and finally broke into two pieces on August 15, 2003. During this period, 70,000 tons of crude oil could have been shifted to other ships, avoiding the environmental disaster.

Dispersants to emulsify oil and minimize the impacts of oil pollution and other environmental hazards are ineffective and again unsafe for coastal communities. The debris and sand mixed with crude oil should be removed and safely disposed of. Oil booms, skimmers, dispersant spray systems, and other response equipment were not available until after the ship became grounded. Oil spill response teams were untrained and unprepared to deal with the crisis.

The geo-politico-economic location of Pakistan has placed it on the oil route of the world. An area of the Arabian Sea is identified as Pakistan's EEZ, and approximately 80% of the world's oil traverses it to be loaded at ports serving the oilfields of Iran, the Arabian Peninsula, and the Gulf States. The tremendous economic advantages have not been shifted to the fisher communities whose habitats and livelihoods are affected by the perils of the sea, including oil spills and shipwrecks.

The federal government did not approve Marine Security Agency's Oil Spill Plan, a proposal prepared in 2000. It had proposed to purchase US $5 million worth of oil spill control equipment, a pollution control vessel at the cost of US $20 million, and some US $10 million for the training of the personnel and the purchase of oil spill dispersing chemicals.

References

Ben Yami M (2002). Safety in small-scale fisheries: What is to be done? In Proceedings of the International Fishing Industry Safety and Health Conference (Woods Hole, Massachusetts, Oct. 23-25, 2000), Lincoln JM, Hudson DS et al., eds. Cincinnati, OH: National Institute for Occupational Safety and Health. DHHS (NIOSH) Pub. No. 2002-147.

Food and Agriculture Organization (Rome) (2000). Selected issues facing fishers and aquaculturists. The state of world fisheries and aquaculture. Parts 1 and 2, pp. 1-64.

Lincoln JM, Hudson DS, and Conway GA (2002). Executive summary. In Proceedings of the International Fishing Industry Safety and Health Conference (Woods Hole, Massachusetts, Oct. 23-25, 2000), Lincoln JM, Hudson DS et al., eds. Cincinnati, OH: National Institute for Occupational Safety and Health. DHHS (NIOSH) Pub. No. 2002-147.

THE NEED FOR LEGISLATION TO IMPROVE SAFETY AT SEA FOR SMALL FISHING VESSELS: THE CASE OF THE PACIFIC ISLAND STATES

Manumatavai Tupou
Kingdom of Tonga
PhD candidate, School of Law, University of Nottingham
United Kingdom

Introduction

For the people of the Pacific Island States, fishing forms an integral part of their culture. The Pacific States covered in this paper are the 14 independent nations of the Cook Islands, Federated States of Micronesia, Fiji, Kiribati, Marshall Islands, Nauru, Niue, Palau, Papua New Guinea, Samoa, Solomon Islands, Tonga, Tuvalu, and Vanuatu. For the purposes of this presentation, "small fishing vessels" refers to fishing vessels under 12 metres in length, motorised or nonmotorised. Fishing by Pacific Islanders is undertaken primarily in these small fishing vessels. It is therefore imperative that adequate regulations are in place to ensure the safety of the users of such vessels. The purpose of this presentation is to provide insight into the status of sea safety laws in the Pacific Island States through the following steps:

Examine the laws that provide safety measures for small fishing vessels.

Highlight the laws that do not provide safety measures for some small fishing vessels.

Suggest ways to incorporate and implement appropriate safety-at-sea legislation for such vessels.

Develop an understanding of the significance of fisheries resources to the Pacific Island States that will substantiate the need for safety measures to protect local fishermen adequately.

Significance of fisheries resources to the Pacific Island States

The geographical composition of these states plainly illustrates the rationale behind the significance they place on fisheries resources. With the exception of Papua New Guinea, these Pacific Island States are predominantly a collection of large Exclusive Economic Zones (EEZs) dotted with small islands.

The EEZs of the Pacific Island States account for 19.8 million square kilometres (Food and Agriculture Organization 1997b). In contrast, their land area amounts to just over 527,000 square kilometres, 88% of which is held by Papua New Guinea (Secretariat of the Pacific Community 2003b). Hence, most of the communities of these states are coastal in nature. The dramatic land-to-sea ratio of these Pacific Island States is clearly exemplified by Kiribati. Kiribati only has 811 square kilometres of land while its EEZ covers 3.55 million square kilometres (Secretariat of the Pacific Community 2003b). Thus, it is no surprise that marine resources form the core of Kiribati's economy. Fishing provides approximately 80% of Kiribati's households with a livelihood (World Bank 2000b).

The foremost significance of oceanic resources to these coastal communities is as a source of food. An estimated 50 kilograms of fish per person is consumed annually (World Bank 2000a). It is also estimated that in the last two decades, the annual per capita consumption is an average of 140 kilograms in Kiribati and an average of 115 kilograms in Tuvalu (Food and Agriculture Organization 2002c, f). This is considered high by world standards, especially when one considers that by comparison the average Australian is estimated to consume 8 kilograms per year (Food and Agriculture Organization 1997a).

As a consequence of their meagre land base, the Pacific Island States have limited land resources. However, these are prone to natural disasters, such as cyclones (most common), earthquakes, landslides, volcanic eruptions, and tsunamis (see note a). Such disasters could lead to several problems, including the destruction of the agricultural sector (see note b.) All these factors intensify the significance placed on fisheries resources.

Small fishing vessels operating in the Pacific Island States

Vessels operating with a Pacific State as a base are (1) motorised and non-motorised local fishing vessels and (2) locally based foreign fishing vessels. Locally based foreign fishing vessels are specifically defined in the laws of the Pacific Island State (see appendix for more information about locally based foreign fishing vessels and related safety concerns). Table 1 clearly illustrates that the majority of vessels in the Pacific Island States consists of motorised and nonmotorised artisanal fishing vessels.

Table 1: Estimates of the number of fishing vessels in the Pacific

Location	Locally based vessels (Gillett 2003)	Motorised artisanal fishing vessels (McCoy 1991)	Nonmotorised arti-sanal fishing vessels (McCoy 1991)
Cook Islands	10L/L	200	120
Fiji	96 L/L; 1 P/L	1,600	400
Fed. States Micro.	34 L/L; 8 P/S	2,000	600
Kiribati	2 L/L; 1 P/S	600	5,000
Marshall Islands	54 L/L; 5 P/S	500	250
Nauru	1 L/L	100	80
Niue	100 skiffs	60	240
Palua	71 L/L; 1 P.S	700	40
PNG	40 L/L; 24 P/S	8,000	10,000
Samoa	153 L/L	80	100
Solomon Islands	8 L/L; 2 P/S; 12 P/L	1,800	5,000
Tonga	26 L/L	800	200
Tuvalu	20 skiffs	200	500
Vanatu	10 skiffs	250	500
Total	495 L/L; 40 P/S; 14 P/L	16,890	24,530

(P/L = Pole and line vessel; P/S = Purse seiner; L/L = Longliner)
Source: R. Gillett, Aspects of Sea Safety in the Fisheries of Pacific Island Countries, Rome: FAO, 2003, 12 (see note c).

The most commonly used local fishing vessels in these Pacific states are canoes and small outboard-powered craft under 12 metres in length. For example, in Niue, the most common crafts are one-man outrigger canoes. The largest fishing vessels are 8.5-metre aluminium catamarans (Food and Agriculture Organization 2002b). In the Marshall Islands, canoes are extensively used for subsistence fishing, while 4.5- to 6-metre craft powered by outboard motors are mainly used in artisanal fishing (Food and Agriculture Organization 2002a).

Research indicates that small fishing vessels are the most vulnerable to safety-at-sea incidents (McCoy 1991; Gillett 2003). An estimated average of one incident of distress per day is reported to search-and-rescue groups (McCoy 1991). Due to communication problems and the remoteness of many islands and villages, this number is likely to be higher (McCoy 1991). Hence, it is critical that appropriate legislation ensures the protection of local fishermen and that this legislation is enforced.

Many of these small fishing vessels are also used for other tasks during which incidents of distress can occur (McCoy 1991; Gillett 2003). The most common alternative use of these vessels is as inter-island transport. For example, in the Federated States of Micronesia in April of this year, four locals transporting supplies from one main island to their home island 60 miles away spent 57 days drifting before they were spotted and brought back to safety (Federated States of Micronesia 2003). Another incident in the Solomon Islands involved a small vessel with five passengers who had gone to a neighbouring main island to do banking. This vessel ended up drifting for 13 days (Pacific Islands Report 2002b). In Fiji, what was meant to be a half-hour return journey on an outboard-powered punt for five villagers following a shopping trip on one of the main islands turned into 13 days adrift at sea. The vessel had no flares or radio and insufficient water and food on board (Pacific Islands Report 2002a). The various uses of small fishing vessels stresses further the need to ensure that safety measures are in place.

Legislation in the Pacific Island States dealing with safety at sea

The relevant legislation containing safety measures for fishing vessels is found in the fisheries and shipping laws of these island states. Apart from Niue, the Pacific Island States generally establish any detail concerning safety measures for vessels in their laws governing shipping. In the majority of these Pacific states fisheries laws, the extent of providing safety for small fishing vessels is to make it a precondition of licensing to (1) comply with relevant national laws governing shipping or navigation in Papua New Guinea, Nauru, Tonga, and the Cook Islands or (2) obtain a seaworthiness or safety certificate under the shipping laws in Kiribati, Samoa, the Solomon Islands, Tuvalu, and Vanuatu. Fiji's and Palau's fisheries legislation does not cover safety at sea. The Federated States of Micronesia and the Marshall Islands fisheries laws broadly state that a license will be denied where the vessel "does not meet required safety standards." However, no such safety standards have been stipulated. Moreover, safety at sea does not appear as an objective of fisheries management in most of the Pacific states. Niue is the only Pacific state that dedicates a part of its Domestic Fishing Act to safety at sea. This reflects Niue's priority to cater expressly to safety at sea for its local fishermen. A possible reason for the other Pacific states not highlighting the profile of safety at sea is that, as developing states, it is common for the objectives of their fisheries laws to focus on maximising economic development and the sustainable use of the fisheries resources (see note d).

With regard to their fisheries and shipping laws, these states can be divided into two categories:

States whose safety measures cover all small fishing vessels and
States whose safety measures cover only some small fishing vessels.

Pacific Island States whose safety measures cover all small fishing vessels

The legislation of Pacific states rarely covers all small fishing vessels. Samoa is an exception to this (see note e). Its regulations apply to all vessels less than 15 metres in length (see Section 2(1), Samoa Shipping (Small Vessels) Regulations 1998.) These regulations, therefore, cover canoes and other small vessels that are used in near-shore fishing for subsistence purposes. The Minister of Transport does, however, have the authority to exempt any vessels from the operation of these regulations or any class of vessels (Section 2(2), Samoa Shipping (Small Vessels) Regulations 1998). These regulations also cover the popular 9- to 10-metre *alias* (aluminium catamarans) powered by outboard motors, as well as catamarans and monohull longliners between 10 to 15 metres long, which operate outside the reefs and offshore.

Initially, *alias* were generally used in bottom fishing inshore (see note f). However, due to the increasingly heavy exploitation of inshore resources, fishermen began seeking resources offshore, especially highly valued tuna. This led to the construction of larger, yet unstable, catamarans. The result was an increased loss of lives at sea, most of which involved catamarans (Secretariat of the Pacific Community 2003b). It is for this reason that the 1998 shipping regulations were introduced. These detailed regulations have made a significant difference to the safety at sea of Samoan fishermen. It would be useful to point out some of the safety measures in these regulations to illustrate the extent of safety measures applied in one of the Pacific Island States.

Under these regulations, fishing vessels in particular must be licensed and hold a seaworthiness certificate, as well as a small vessel safety certificate (see Section 3(1)-(2), Samoa Shipping (Small Vessels) Regulations 1998). These certificates are issued by the Secretary of Transport. An application for such certificates must be accompanied by another certificate from the Department of Labour which affirms the vessel's compliance with occupational, safety, and health requirements (see Section 3(3), Samoa Shipping (Small Vessels) Regulations 1998). A fishing certificate may be refused by the Secretary of Transport when the vessel

- Does not meet the requirements in these regulations.
- Is not seaworthy.
- Is not equipped with the requisite safety equipment as stated in these regulations.
- Is not constructed by authorised vessel builders.
- Is, in the Secretary's opinion, a threat to life, property, or the environment (see Section 3(4), Samoa Shipping (Small Vessels) Regulations 1998).

The specifications and construction plans to build a vessel must be submitted to the Secretary of Transport for his approval (see Section 5(1), Samoa Shipping (Small Vessels) Regulations 1998). Also, small fishing vessels operating within 20 nautical miles offshore are required to carry specific safety equipment, including the following:

Lifejackets for every person on board.
Appropriate marine compass.
Storm (sea) anchors.
Parachute rockets, flares, and smoke signal.
VHF marine radio with the appropriate channel.

Engine spare parts and tools for repairs at sea.
First-aid kit.
Fire extinguisher.
Survival rations or food and potable water for all persons on board, sufficient for at least 3 days.
Waterproof handheld torch or mounted searchlight.

The regulations also provide specific requirements for the manning, training, and certification of crew (see Section 7, Samoa Shipping (Small Vessels) Regulations 1998).

The life-saving changes resulting from these regulations can be clearly illustrated in Table 2. This table shows that the number of fatalities at sea significantly dropped and continually decreased after the 1998 introduction and active implementation of these regulations.

Table 2: Statistics of Samoan fishermen lost at sea

Year	Number of fishermen lost at sea
1988	22
1997	17
1998	7
1999	6
2000	5
2001	3
2002	0

Source: Secretariat for the Pacific Community (2003a), 'A Samoan success story', in Regional Maritime Programme Newsletter Issue 19 (March 2003), 3; the 1988 figure was provided by F. Lafoai, Head of School of Maritime Training, Samoa (11 September 2003).

The success of these regulations in minimising, and in 2002 eliminating, the loss of lives at sea is attributed to the regulations, discussions with stakeholders, and extensive follow-up work in the form of a training and education programme (Secretariat of the Pacific Community 2003a). In this regard, Samoa's regulations and the implementation process that followed are highly commended and provide a positive example for other Pacific Island States.

States whose safety at sea measures cover only some small fishing vessels

Most of the Pacific Island States fisheries and shipping laws only apply to certain types of small fishing vessels. The most common type of exemption in these states' laws is vessels under a specific length. For example, Tonga's Fisheries Act does not apply to vessels under 6 metres, and its Shipping Act does not apply to vessels under 8 metres. Vanuatu's Fisheries Act does not cover fishing vessels under 10 metres, and its Shipping Act excludes boats whose sole means of propulsion is through manpower. Similar exemptions are made in most of the other Pacific Island States. In doing so, these laws exclude numerous small fishing vessels from their scope of regulation and, as a consequence, only require some small fishing vessels to carry safety equipment. In Kiribati and Tuvalu, such exemptions result in the exclusion of most, if not all, privately owned fishing vessels, which are used for inshore and offshore fishing for commercial and subsistence purposes (see note g).

Another common method of exempting some small fishing vessels in these states' laws is to exclude those small vessels, often canoes, that fish on a noncommercial basis. For example, the Solomon Islands Fisheries Act and Shipping Act only cover vessels that fish on a commercial basis, and Niue exempts canoes from its Domestic Fishing Act and subsidiary legislation. This is of critical concern, considering that most local fishing that occurs in the Pacific Island States is for noncommercial or subsistence purposes (World Bank 1996; Food and Agriculture Organization 1997a, b). There is no rationale for making a craft used for commercial purposes safer than a craft used solely for subsistence means. Exempting such vessels could prove perilous for subsistence fishers. For example, by exempting canoes from its domestic fisheries laws, Niue excludes the most common craft used for subsistence fishing. This lacuna in the law is underlined further by there being 200 canoes as compared to 62 registered local fishing vessels (Food and Agriculture Organization 1997b). Also, in the late 1990s in Niue, two of the four incidents resulting in fatalities at sea involved canoes (the other two involved diving) (Secretariat of the Pacific Community 1999).

An obvious reason for the exemption of smaller fishing vessels and canoes from the scope of most of these states' laws was highlighted in a 1991 FAO study. This study found that officials in most of the Pacific states generally believe that it would prove impossible to enforce any safety regulations on such vessels (McCoy 1991). Most Pacific states consist of several islands that are remote from the main islands, resulting in communication and enforcement problems. Another reason for the exemption could be that small fishing vessels tend to fish close to shore and are less likely to be involved in sea safety incidents. This deduction overlooks the fact that small fishing vessels continue to venture offshore to fish, not just for commercial purposes, but also for subsistence means, e.g., in Tonga and Tuvalu (Preston, Chapman et al. 1987; Gillett 2003).

It has been established that there are threats of overfishing in some Pacific states' coastal areas and lagoons (World Bank 2000b). Coupled with the cash economy, this may entice more users of small fishing vessels to pursue fishing offshore and thereby carries implications for their safety, as was the case in Samoa.

Suggestions on ways to improve safety at sea for users of small fishing vessels

The following suggestions could assist Pacific Island States in taking steps toward promoting and improving the safety of local fishermen. These suggestions are grouped into three steps:

1. Groundwork for any changes to the law,
2. Establishing safety laws, and
3. Implementing and enforcing such laws.

Groundwork for any changes to the law

1. A prerequisite to amending any of these Pacific states' laws is extensive consultation with individual fishing communities regarding existing small fishing vessels and equipment and the scope of their use.

2. Data on sea-safety incidents must be collected and documented to establish an informed view on which vessels are vulnerable at sea (Gillett 2003). In the villages or remote islands, much of this information will be based on oral information. These data could also be collected from the relevant government authorities responsible for search-and-rescue operations.

3. Such information allows any consequential amendments to the law to be adapted to fishers' real-life situations to improve their safety at sea and complement, rather than radically change, their existing fishing customs. This would also promote their compliance with such laws.

4.   Local officers already based in fisheries divisions in the outer islands should be employed to facilitate such meetings. Meetings in villages could be conducted through the use of local authorities and traditional village gatherings such as a *fono* in Tonga.

5.   Based on the findings of such consultations, any small fishing vessel considered to be vulnerable to safety-at-sea incidents must be included in the scope of application of the fisheries or shipping laws.

6.   It is critical that to change the law in any of these island states, political support must be sought to highlight the need for any safety-at-sea measures (Gillett 2003). This could be encouraged through a workshop to which national and local leaders are invited. Information on the extent of safety-at-sea incidents and the cost to government of search-and-rescue operations could be presented at this workshop. This could prompt political leaders to direct any necessary funds for search-and-rescue operations.

7.   The extent of safety-at-sea incidents should be publicised. This will contribute to highlighting safety at sea for local fishermen at a national level.

Establishing safety laws

8.   Any subsequent actions resulting from this information must still involve local input. In Pacific Island States with outer islands and remote communities, it is critical that fisheries officers continue to work together with village leaders, as well as the fishermen themselves, as they could assist in tailoring any rules more closely to community needs (World Bank 2000b).

9.   The Regional Maritime Programme of the Secretariat of the Pacific Community has established some model maritime regulations that should be consulted when drafting safety-at-sea measures. Such model regulations can be tailored to each states' particular needs. Some Pacific states are already using these model laws as a basis, such as Palau, which currently has an Admiralty and Maritime Bill before its Congress (Moses 2003).

10. Any safety rules must be kept simple. It has been found that "simple management rules work best" (World Bank 2000a). The simpler the rule, the more easily understood and followed.

11. Domestic legislation that is seen to be relevant to the community and endorsed by village leaders as local rules achieves the most compliance (World Bank 2000b). It is, therefore, imperative that local awareness programmes target village leaders to promote compliance with safety at

sea measures.

12. Safety at sea needs to feature prominently in the Pacific Island States' fisheries laws. One of the obvious ways to do this is by incorporating safety at sea as a fisheries management objective, not only in the principal acts governing fishing, but also as one of the priorities in the work agenda (Turner and Petursdottir 2002; Gillett 2003). This will serve to elevate the status of safety at sea and encourage the channelling of resources to this significant area.

Implementing and enforcing safety laws

13. The role of enforcement is critical to the implementation of the law; otherwise, the law would be futile. The lack of enforcement capacity in most of these island states poses a problem. However, community leaders and members may assist in enforcement. For example, in Fiji, the Minister for Fisheries may appoint honorary fisheries wardens (see Section 3, Fiji Fisheries Act 1992), and in the Cook Islands, local fisheries committees and the island councils could assist (see note h). Such island councils can also be found in Kiribati (Food and Agriculture Organization 2002c). In the Marshall Islands, local governments are responsible for managing their coastal fisheries within 5 miles (see Section 43, Marshall Islands Marine Resources Act 1997). In Palau, each of its 16 states is delegated responsibility over its territorial waters (12 miles from shore) (see Section 2, Constitution of the Republic of Palau).

14. Wherever possible, it is also important to make use of traditional systems of authority within these coastal communities when implementing any new sea safety measure. The employment of such systems has proven to be successful for fisheries management initiatives in Samoa. Customary fisheries management in the Solomon Islands has also been effective (Food and Agriculture Organization 2002e).

15. The assistance of fishermen's associations should also be enlisted in drawing up any safety-at-sea measures. The involvement of such associations in this exercise is another means for publicising such measures to the fisher community. A useful strategy would be for fishermen to refuse to go to sea on unsafe vessels (Gillett 2003). Fishermen could, therefore, play a vital role in ensuring that boat owners comply with the law.

16. Effective deterrents for violations have proven to promote compliance with management rules. Governments should support "collaborative enforcement" with lower levels of government and local village leaders.

The publication of successful prosecutions and associated penalties will assist in promoting compliance (World Bank 2000a).

17. It is crucial to raise public awareness for safety at sea through appropriate media.

   a. This can include news presenters on radio and television. Television and videos are most commonly found in urban areas, which is where most commercial fishing takes place.

   b. Such presentations could highlight the human cost of failure to comply with the law through publicising actual safety-at-sea incidents or deliberate flouting of safety measures. For example, in Samoa a tragic incident in 2001 highlighted the need for government officials and boat owners alike to adhere to the safety regulations that had been enacted to protect local fishermen. An *alia* owner was granted a fitness certificate despite clear knowledge by both the officials and the boat owners that the *alia* was unfit to go to sea. Before setting off on the fishing trip, the fishermen had alerted the boat owners that a spare spark plug for the engine was needed, yet none was provided. The vessel also did not have any personal flotation devices or flares. The spare engine did not work, and the radio had no battery. Following engine failure, the vessel ended up drifting for 5 weeks. Only two of the four fishermen survived (Pacific Islands Development Programme East/West Centre 2001a, b).

   c. Such awareness campaigns should be presented in the official or national language and wherever possible in the language or dialect of relevant communities.

   d. This awareness campaign could target and utilise local authorities (often the elders in the village), women, children, and schoolteachers. For example, in Nauru, school children participated in a maritime safety project to make them more aware of vessel safety requirements (Pacific Islands Development Programme East/West Centre 1998). The family and community networks within these Pacific communities can have positive implications for promoting safety-at-sea measures and assisting in their implementation.

18. It is also very important to support the safety at sea programmes in the various marine or maritime centres situated in many of these Pacific states.

19. Training within relevant government administrations is critical. This can occur on several levels.

   a. The training of relevant government officers on the significance of safety at sea. This should include those responsible for search and rescue to ensure they document safety-at-sea incidents thoroughly. In some states, such operations may be undertaken by one authority, such as the Navy in Fiji, whereas in states such as Tonga, it involves the Ministry of Police, the Ministry of Marine and Ports, and the Tonga Defence Services.

b. Involving fisheries officers in implementing safety at sea programmes at a village level;

c. The appointment of a fisheries officer at the head office whose role is to oversee the implementation of safety-at-sea requirements and act as the contact point for difficulties in implementation by local fisheries officers in the outer islands and local authorities.

d. Developing long-term strategies within the mandates of fisheries' administrations to reinforce any such safety-at-sea measures (see note i). The financial situation of the Pacific state, especially that of the local fisherman when they need to obtain safety equipment, is a critical consideration in establishing any safety measures.

In conclusion, it is clear to see that the above suggestions need to be implemented in steps by individual Pacific Island States as short-term and long-term work strategies. These suggestions should prove useful when establishing appropriate legislation to improve safety at sea for the users of small fishing vessels in the Pacific Island States.

Acknowledgments

The author expressly thanks those who kindly assisted by responding to some queries relating to this paper, namely Jeremy Turner (FAO), Josie Tamate (FFA), Peter Heathcote (SPC), Ned Howard (Cook Islands), Mathew Chigiyal (FSM), Glen Joseph (Marshall Islands), Peleni Talagi (Niue), William Moses (Palau), Fatu Lafoai (Samoa), Tevita Poasi Tupou, Gordon and Meg Ward (Tonga/Solomon Islands), and Peter Murgatroyd (Vanuatu).

References

Federated States of Micronesia (2003). Chuuk foursome survives six weeks at sea. News and Public Statements page, Federated States of Micronesia.

Food and Agriculture Organization, Fisheries Department (1997a). Fisheries and aquaculture in the South Pacific: Situation and outlook in 1996. FAO Fisheries Circular No. 907 FIP/C907.

Food and Agriculture Organization, Fisheries Department (1997b). Review of the state of the world fishery resources. Marine Fisheries 16. South Pacific Islands. Fisheries Circular No.920 FIRM/C920.

Food and Agriculture Organization, Fisheries Department (2002a). Fishery country profile: Marshall Islands.

Food and Agriculture Organization, Fisheries Department (2002b). Fishery country profile: Niue.

Food and Agriculture Organization, Fisheries Department (2002c). Information on fisheries management in Kiribati.

Food and Agriculture Organization, Fisheries Department (2002d). Information on fisheries management in the Solomon Islands.

Food and Agriculture Organization, Fisheries Department (2002e). Information on fisheries management in Tuvalu.

Gillett R (2003). Aspects of sea safety in the fisheries of Pacific Island countries. Rome: Fisheries Department, Food and Agriculture Organization.

Gillett R, McCoy M et al. (2001). Tuna: A key economic resource in the Pacific Islands. A report prepared for the Asian Development Bank and the Forum Fisheries Agency. Manila: Asian Development Bank.

Moses W (2003). Personal communication with W. Moses, manager of Palau's Maritime Safety Branch, Division of Transportation and Communication.

McCoy M (1991). Survey of safety at sea in Pacific Island artisanal fisheries. F. D. 91/3. Suva. FAO/UNDP Regional Fisheries Support Programme.

Pacific Islands Development Programme East/West Centre (1998). Nauru offers marine safety program to students.

Pacific Islands Development Programme East/West Centre (2001a). Fishing Boat accident survivors welcomed home to Samoa.

Pacific Islands Development Programme East/West Centre (2001b). Slack and ignorant: Samoa fishing boat owners attitude alarms.

Pacific Islands Development Programme East/West Centre (2003). PNG maritime licences sold under counter.

Pacific Islands Report (2002a). Five Fiji villagers survive 13-day ordeal at sea. Pacific Islands Development Programme.

Pacific Islands Report (2002b). Five Solomon Islanders survive 13 days adrift. Pacific Islands Development Programme.

Preston G, Chapman L et al. (1987). Trolling techniques for the Pacific Islands: A manual for fishermen. Secretariat for Pacific Development.

Secretariat of the Pacific Community (1999). First SPC heads of fisheries meeting country statement–Niue. Information Paper 24. Noumea New Caledonia.

Secretariat of the Pacific Community (2003a). A Samoan success story. Regional Maritime Programme Newsletter.

Secretariat of the Pacific Community (2003b). Selected economies: A statistical summary.

Talagi P (2003). Personal communication with P. Talagi, Crown Counsel, Niue.

Tamate J (2003). Personal communication with J. Tamate, project officer, Forum Fisheries Agency, Solomon Islands.

Turner J and Petursdottir G (2002). Safety at sea for fishermen and the role of FAO. In Proceedings of the International Fishing Industry Safety and Health Conference (Woods Hole, Massachusetts, Oct. 23-25, 2000), Lincoln JM, Hudson DS et al., eds. Cincinnati, OH: National Institute for Occupational Safety and Health. DHHS (NIOSH) Pub. No. 2002-147.

World Bank (1996). Pacific Island Economies: Building a resilient economic base for the twenty-first century. A World Bank Country Study. Washington DC.

World Bank (2000a). Meeting challenges in the global economy: Report of the Commonwealth Secretariat/World Bank Joint Task Force on Small States. Washington DC.

World Bank (2000b). Summary report: Voices from the village. A comparative study of coastal resource management in the Pacific Islands.

Notes

a. Vanuatu alone recorded 29 cyclones between 1970 and 1985. A similar frequency of tropical cyclones is experienced by most of the other Pacific Island States. The common occurrence of natural disasters in the region is a grave concern as it severely hampers and disrupts economic and social development due to the high dependence of islanders on their natural environment (South Pacific Regional Environment Programme, South Pacific Programme Office United Nations Department of Humanitarian Affairs and Emergency Management Australia, *Natural Disaster Reduction in Pacific Island Countries Report to the World Conference on Natural Disaster Reduction: Report to the World Conference on Natural Disaster Reduction Yokohama, Japan 23-17 May 1994*, Australia: Emergency Management Australia, 1994, 5).

b. Other problems include villages being wiped out, severe interruption to communication services and significant numbers of human fatalities (F. K Tevi, *Vulnerability: A Pacific Reality*, Suva: Pacific Concerns Resource Centre, 6).

c. In June 2001, Niue had 200 canoes and 62 registered boats. The total does not include skiffs; the number of skiffs was only noted in countries where industrial tuna vessels are absent.

d. Although safety-at-sea provisions may not feature prominently in most Pacific States fisheries laws, in general these same laws allow a recognised authority such as the Minister of Fisheries to enact any necessary regulations. Fisheries legislation often lists areas that may be regulated. Some laws specifically list one of these areas as safety measures for local fishermen and fishing vessels (for example in the Cook Islands, the Solomon Islands, and Tonga). These lists are nonexclusive. Hence, if safety measures are not expressly stated, the recognised authority can still stipulate such regulations.

e. Nauru provides a second example of a state that covers all small fishing vessels in the scope of its legislation. Nauru's Fisheries Act 1997 covers small boats not more than 10 metres in length that are used or intended to be used solely in coastal or internal waters. Under Nauru's Fisheries Regulations 1997, small boats must be registered with the Ministry of Fisheries. The Chief Executive Officer of the Nauru Fisheries and Marine Resources Authority may register a small boat where he is satisfied, amongst other things, that the boat meets "acceptable safety standards" (standards required from time to time by the board of directors for the Authority or the Chief Executive Officer). One of the safety requirements is that the boat is fit for fishing or that the boat carry a valid sea-going certificate or seaworthiness certificate as required under the laws governing shipping (section 15(2)(d) Fisheries Regulations 1997).

f. The *alias* were a FAO design introduced in the mid-1970s (FAO, Fishery Country Profile: Samoa, Rome: FAO Fisheries Department, 4).

g. Kiribati's Fisheries Act exempts vessels under 7 metres from the scope of its application. However, the vast majority of its offshore fishing (for commercial and subsistence purposes) is undertaken in vessels less than 7 metres (M. Miria-Tairea, *Fisheries Legislative Profile: Republic of Kiribati*, FFA Report 95/2, Honiara: FFA, 1995, 13). Similarly Tuvalu's Fisheries Act excludes vessels under 7 metres, yet there are no privately owned vessels over 7 metres operating in Tuvalu (Gillett, *supra* n.14, 20). On Tuvalu's main island Funafuti, 4-to 5-metre outboard motored skiffs are mainly used in trolling for tuna (FAO, Country Profile: Tuvalu, Rome: FAO Fisheries Department, April 2002, 5).

h. Section 4-7 Cook Islands Marine Resources Act 1989. On one of its islands, Aitutaki island, a fisheries management initiative to establish marine protected areas failed due to the initial lack of community support. However, with the support of the traditional leaders, the Aitutaki Island Council, and the community, this initiative was eventually established. This underlines the significance in engaging traditional and local leaders support for fisheries management (Government of the Cook Islands, *Ministry of Marine Resources Annual Report for the Year ended 30 June 2000*, Cook Islands: Ministry of Marine Resources, 2000, 16). The responsibility of enforcing management measures lies with the Aitutaki Local Council (FAO, Information on Fisheries Management in the Cook Islands, Rome: FAO Fisheries Department, April 2002, 5).

i. The need for continual reinforcement of fishing safety management initiatives was highlighted in McCoy, *supra* n. 14, 61.

j. The 495 locally based vessels in 2003 do not include these skiffs.

Appendix : Locally Based Foreign Fishing Vessels and Some Safety Aspects

Definition

These vessels are generally defined in the Pacific Island States' laws as foreign vessels that are "based" in one of these Pacific states. Some of the island states vary as to the meaning of "base." For example, in the Cook Islands, this refers to a foreign vessel based in the Cook Islands or a foreign vessel based in another Pacific state and jointly operated by or on behalf of the Cook Islands' government and one or more other governments in the South Pacific region, under an agreement to which the Cook Islands is a party (see Cook Islands Marine Resources Act 1989, s.2). In Samoa, the vessel has to be based there and land "any of its catch in Samoa" (see Samoa Fisheries Amendment Act 1999, s.2). Similar to Vanuatu, Tonga defines such vessels as a foreign vessel "based in Tonga and landing all of its catch in Tonga" (see Vanuatu's Fisheries Act 1982 (Rev. Ed 1988), s. 1; Tonga Fisheries Act 1989, s. 2; Tonga's Management Bill 2002, s. 2.).

In mid-2000, there were 304 locally based foreign fishing vessels (above 15 metres in length) operating in the Pacific Island States. Most of these vessels were based in Palau (81) (Gillett, McCoy et al. 2001). In 2003, there were 549 locally based foreign fishing vessels, 495 of which are longliners. Most of these longliners are based in Fiji (96) and Palau (77) (Gillett, McCoy et al. 2001). Locally based foreign fishing vessels are mainly comprised of longliners, purse seiners, and pole and line fishing vessels. Skiffs from other states also operate as locally based foreign fishing vessels in Niue, Tuvalu, and Vanuatu (see note j).

Specific safety aspects relating to locally based foreign fishing vessels

1. Recently, Niue entered into a joint venture agreement with Samoa that entails 100 locally based skiffs fishing in Niuean waters (Tamate 2003). Niue's Domestic Fishing Act contains detailed safety provisions for its motorised small fishing vessels. This act also covers a commercial cargo vessel having a gross tonnage "less than five tonnes." A vessel, for the purposes of the act, refers to any boat, aircraft, ship, or other sea-going craft. Hence, these locally based foreign skiffs are covered by this act and its subsidiary legislation. It is imperative that the safety measures in Niue's laws are enforced on these skiffs so that the rise in sea safety incidents that occurred in Samoa (due to the pur-

suit of highly prized tuna and lack of safety regulations) is not repeated here.

2. Having said this, enforcement capacity in Niue, as in the other Pacific Island States, is very limited (Talagi 2003).

3. Some suggestions that have been made for local fishing vessels in the presentation can be applied here. For example, it is imperative that relevant government personnel fully appreciate the significance of safety at sea for seafarers through training. This is a crucial step toward promoting sea safety. Such training will help to avoid problems such as the sale of ship engineers' certificates to unqualified people by government officials. This transpired in Papua New Guinea and resulted in several safety incidents (Pacific Islands Development Programme East/West Centre 2003).

4. Some of the Pacific Island States receive an important source of revenue from their nationals working as seafarers on locally based foreign fishing vessels and foreign fishing vessels. Hence, maritime training programmes within these states and at the University of the South Pacific must be fully supported to ensure that the quality of training is in accordance with internationally accepted maritime standards. It is also imperative that such states provide adequate sea safety measures within their relevant laws to protect their nationals working on such vessels and ensure these laws are enforced. Sea safety measures can be made conditions of licensing or as terms of fishing access agreements.

CHILEAN SAFETY ASSOCIATION AND THE FISHING SECTOR

Eduardo Rossel
National Coordinator Fisheries Program
Chilean Safety Association
Santiago, Chile
E-mail: erossel@achs.cl

First, I would like to express my gratitude for this opportunity to share the Chilean industrial fishing sector's experience with maritime safety.

At present, the Chilean Safety Association, with its 37,000 associated companies, oversees all the economic activities of approximately 1.5 million affiliated workers.

The Chilean Safety Association–Asociación Chilena de Seguridad (ACHS), which I represent–administers the Bill of Work of Accidents and Professional Illnesses. In Chile, three entities provide risk prevention, medical care, economic compensation, and vocational rehabilitation to their affiliated companies and workers. ACHS, the largest, represents 53% of the industry's workers.

The fishing industry in Chile ranks third in total exports for the country, representing 12%, immediately behind the mining and fruit industries.

The technological advancements within the Chilean fishing industry have led to more automated methods for harvesting and processing the resource. This implies the industry is operating more efficiently, but work tasks and risks have changed, leading to changes in crew requirements, continuous learning, and specified training.

Since 1992, fishing companies associated with ACHS have experienced fewer accidents, although the accident rate within the fishing industry still exceeds the accident rates associated with other industries. It is estimated that accidents in the fishing sector cause our country direct and indirect losses of around US $35 million a year.

Globalization and technological changes have produced enormous challenges that have resulted in important advances for Chile's industrial fishing sector. Because of modernization, fishing requires less physical exertion and brute strength, and subsequent exhaustion of crew members has been lessened. The accident rate, in turn, has decreased. Innovations in fishmeal plants have resulted in a more environmentally friendly industry, and investments in workers training and working conditions have led to improved operations and better worker performance.

The Chilean fishing sector is primarily made up of purse seiners (70%), trawlers (19%), and longliners (11%). The principal resource is pelagic fisheries. The principal fishing centers in Chile are located in the 1th Region (Arica and Iquique–northern area) and in the 8th Region (Concepción and Talcahuano–central southern area), accounting for 71% of the national fish landings. The fishing industry is made up of about 60 thousand workers directly related to fishing. During 2002, the fishing sector showed the greatest number of new jobs, with a 9% increase.

The Chilean fishing fleet is made up of 410 vessels, which caught more than 4.8 million tons in 2002. Forty percent of the vessels in the Chilean fishing fleet range from 101 to 400 gross registered tons (GRT), 50% of its vessels are under 25 years old, and 49% of these boats were built in Chile.

The work performed by Chilean fishermen is considered "heavy work." That is to say, "it is the kind of work which most rapidly damages workers physically, psychologically, and intellectually, causing precocious aging, though not generating professional diseases." This definition of heavy work was established by an Ergonomics Commission formed by representatives of the Chilean government, companies, and workers. The commission makes clear the interest of the Chilean legislation to protect and provide for the fishing vessel crews.

Regarding safety matters and the high occupational accident rate occurring in the whole sector, ACHS has developed effective safety programs to control risks in fishing companies, especially those taking place on board fishing vessels. Our indices indicate that 69% of the industry's accidents occur in fishing fleets and 31% in fishmeal processing plants.

Within the framework of support to companies associated with ACHS, our specialists have developed programs to assist the fishing sector, which resulted in a decrease of the accident indices from 24% in 1992 to 14% in 2002. We believe that this accident reduction is associated with the implementation of our plans, programs, and preventive measures, especially emphasizing the work on board commercial fishing vessels.

The incorporation of technological advances into fishing methods and practices has brought about other risks that must be made known to the crews. The working conditions on board fishing vessels have been improved, as has the quality of life, mainly in the areas of comfort and habitability. Companies and crews have understood that advances in accident prevention not only make work safer, but also allow the Chilean fishing industry to be more competitive with other fishing countries of the world.

In relation to factors associated with accidents occurring on board fishing vessels belonging to the companies associated with ACHS, an analysis of approximately 6,000 cases revealed the following:

+ The crew members involved had been working at the company more than a year, and 62% had more than 1 year of experience with the job in which they were injured.
+ Twenty-nine percent of accidents were classified "hit by/against objects"; "overexertion" accounted for 23%.
+ Thirty-five percent of accidents resulted from moving or maneuvering objects.
+ Hands and feet were the most frequently injured body areas, accounting for 19% and 18% of the accidents, respectively.

From the information obtained through the ACHS accident research and from other related statistics already in existence, accidents occurring to workers in the Chilean fishing industry result primarily from incorrect or deficient working methods and poor risk prevention training on both the management and worker levels. Other factors that may lead to fishing-related accidents include workers' social and cultural characteristics; not taking safety training seriously, partially because of the implementation of more automation; lack of injury-prevention-related policies in companies; and finally, working atmosphere and environmental conditions, including rough seas and night work. Safety is not considered to be an isolated and specific problem, but an integral part of the administration and efficiency of a company. Active participation of all the people belonging to the organization is required, whatever their position in a company's hierarchy. This matter has significant importance on fishing vessels. It is urgent to change some paradigms and assumptions existing among the crews, such as "This work has always been done this way," "There is no other choice, so it has to be done," "The vessels cannot stop," or "Performance is first"–all assumptions that have delayed the full development of safe work on board fishing vessels.

The strategic plan developed by ACHS for the fishing sector has been directed to those areas still presenting a high accident rate. Training is mainly focused on the risks associated with the tasks performed on board the vessels or on the training required by the maritime authorities in relation to International Maritime Organization (IMO) courses. Primary focus and consideration is given to the following:

+ Diagnosing the fishing companies' plants and fleet in order to know the degree of development of the Risk Prevention Administration, determination of its strengths and weaknesses, defining strategies that permit operations with acceptable risk levels.
+ Obtaining written documentation of safety policies and how they are applied from company management.
+ Obtaining descriptions of injury prevention programs from fishing companies.
+ Training in the risks associated with industrial fishing, specifically maneuvering, casting of fishing nets, legal issues, and vessel rules.
+ Acknowledging risks on fishing vessels reporting on-board conditions.
+ Enforcement of SOLAS rules, supervised by the Chilean maritime authorities (life-saving equipment, IMO model courses).
+ Performing emergency drills at reasonable time interval.
+ Performing risk evaluations, principally in vessel machine rooms and taking noise levels into consideration.
+ Forming a close relationship between ACHS and the maritime authorities in matters such as rules, courses, and certifications.
+ Supporting crews by publishing information related to maritime safety, such as manuals and booklets, as recommended by the International Labour Organization.
+ Producing videos about sea survival and maritime safety for the different fishing techniques used.

Finally, I would like to point out that to attain better safety levels on fishing vessels, it is necessary to increase the learning and training activities within the fleets, actions that must be carried out using simple, easy-to-understand vocabulary. Therefore, it should be understood that prevention is a continuous improvement task process based on a process of cultural changes within companies to include a safety culture.

ACHS has entered an agreement between our institution and an important insurance agency of Perú. The growth of the fishing industry in Perú is relatively new, and Chile has had a law of work accidents from 1968. Our experience with the subjects of safety and prevention of risk are very solid. This sharing agreement allows the ACHS to give from the experience and knowledge of more than 40 years.

The size of the fishing vessels and the type of fishing methods used by Peruvian fleets are very similar to those of Chile. Between Perú and Chile, an important technological interchange exists, as much in equipment for the flour plants as for fishing vessels. Let us remember that both countries are among the top five fish flour producers in the world. Perú and Chile have needed to develop and maintain a permanent interchange to allow the fishing companies of the region to produce with quality in accordance with environmental exigencies and within a safety culture. The consultant's office of the ACHS is charged mainly with visiting the fishing companies of Perú and fish flour plants, processors, and fleets in their different ports–Paita, Chimbote, Pisco, and Ilo. During these visits, risk analyses are made, and injury prevention and marine safety training is offered. One difference between the expansion of safe practices and safety training in Chile versus Perú is that in Chile, laws addressing safety in the fishing industry and protection of its workers apply to all situations and all workers within the industry–whether in processing plants or at sea. In Perú, current law only applies to the fishing fleets. Hopefully, this will be corrected.

I would like to express my deep gratitude for this opportunity to present and describe some matters related to fishing activity in Chile. I am certain that conferences like this are magnificent opportunities to exchange experiences and definitely allow the fishing sector to develop new policies and incorporate actions to promote risk prevention among fishermen with greater energy. If that is possible, we can reduce the number of maritime accidents throughout the world, including South America.

RISK ASSESSMENT (SAFETY MANAGEMENT) IN THE UK FISHING INDUSTRY

Captain Clifford W. Brand, BSc (Hon's), DipMarSar, MNI
Inspector of Marine Accidents
Marine Accident Investigation Branch
United Kingdom
E-mail: cliff.brand@dft.gsi.gov.uk

Introduction

The UK Merchant Shipping and Fishing Vessels (Health and Safety at Work) Regulations 1997, which came into force on 31 March 1998, requires all employers (fishing vessel owners and operators) to adopt a risk-based approach to occupational health and safety.

The regulations require fishermen to carry out a risk assessment of the health and safety of workers arising in the normal course of their activity.

This paper focuses attention on the concept of risk assessment in the UK fishing industry and the benefits of such an approach. However, it also looks at the general lack of compliance by the industry, the concern of UK regulators and safety organisations, and the way ahead.

Background

The UK fishing industry is very diverse and embraces a wide range of activities, from deep water pelagic fishing to scalloping, and from coastal netting to single-handed potting. In June 2002, there were 7,045 UK registered fishing vessels.

Although fishing's contribution to the UK's Gross Domestic Product is small, it is a major employer in some coastal communities and an established way of life for families who have been in the business for generations. The culture of the industry is steeped in the past, and those engaged in it are forever preoccupied with the traditional problems of battling with the elements, dwindling stocks, ever more regulations, what they see as unsympathetic bureaucrats, unrealistic quotas, and ever-rising costs.

Although many UK fishermen have an instinctive feel for safety and practice it in their own way, others virtually ignore it. Given a choice, they will focus any new investment into other areas, especially if it improves their ability to catch fish.

Although many fishermen will maintain they are safety-conscious, the evidence in the UK from vessels involved in accidents strongly suggests otherwise. There is no doubt many fishermen find that maintaining their vessels and safety equipment in good order is extremely expensive. Costs weigh heavily on their minds, and they will argue that meeting any new regulation to improve safety is prohibitively expensive. But as the UK Marine Accident Investigation Branch (MAIB) frequently points out, nearly all the fishing vessel accidents investigated by the Branch could have been prevented, not by investing in large sums of money, but by exercising greater care. Again and again it has been found that the enemy of safety is not so much a shortage of money as it is the fishing industry's failure to adopt a safety culture, with everyone doing their best to prevent accidents from happening in the first place.

Not wanting to impose more regulations on the UK fishing industry, particularly in light of the costs involved, a risk-based approach was advocated as a way forward to reduce the number of accidents in the industry.

Method

During the course of the author's work as an inspector with the MAIB, several accidents that resulted in fatalities were investigated where risk assessment has been, or should have been, undertaken to identify the need for preventive measures. These investigations included several visits to fishing vessels in service and exhaustive interviews with skippers and crews.

Health and safety at work regulations
The current UK Merchant Shipping and Fishing Vessels (Health and Safety at Work) Regulations (SI 1997/2962) have been in effect since 31 March 1998 and are promulgated in UK Marine Guidance Note 20. These regulations place duties on all *employers* and *workers* on board ships. A worker is

defined as "any person employed by an employer under a contract of employment, including trainees and apprentices." There are no exemptions.

All employers have a duty under these regulations to ensure the health and safety of workers and others affected by their activities so far as is reasonably practicable. The basis of all safety measures taken should be an assessment by employers of any risks to workers' health and safety from their work activities, commonly known as a health and safety risk assessment. This assessment does not have to be written. However, it is made clear in the available advice relating to risk assessment that it is beneficial do to so, especially if litigation is involved.

Risk assessment

Risk assessment is not a new concept; it has been common practice and effective in industry ashore for many years. It has also been used at sea over the years, albeit in a less formal manner. Since its use on board ships was made mandatory, however, accident investigations have revealed that compliance has been sporadic. During the last 3 years, since this conference was last held in 2000, a further 57 lives have been lost from fishing vessels. This safety record is inevitably open for improvement, and a more widespread use of risk assessment has the potential to contribute in this respect.

The lack of employment contracts through single-handed operation or partnerships theoretically excludes some operators from compliance with the regulations, as they are not deemed to be workers in so much as the regulations are concerned. This issue currently is being tested in the UK courts. It is not known how many fall into this category, but considering there are approximately 6,037 small commercial registered fishing vessels (under 15-m registered length) operating in the UK, there will be many that do. It is more than likely that the operators of many of these vessels have not conducted a risk assessment because they are either unaware of the requirement to do so or do not understand the process. This has certainly been the case in the majority of fishing vessel accidents investigated by the author.

Advice in the UK for carrying out a risk assessment is promulgated in a Marine Guidance Note (MGN 20), which a majority of operators might not have read. Consequently, they might not be aware of the requirement to conduct a risk assessment. Operators that have read MGN 20 could be forgiven if they were baffled by its contents. Its style is very formal, and the guidance provided on risk assessment in its annex and appendix appears very technical. Persons unfamiliar with risk assessment could find this guidance off-putting and difficult to put into practice. As compliance with the health and safety regulations is not usually audited, other than when a serious accident has occurred, the failure to conduct a risk assessment, for whatever reason, frequently goes unnoticed. Because risk assessments do not have to be written, such an audit would be extremely problematic.

To help fishermen with the process of completing a written risk assessment, the Sea Fish Industry Authority (SFIA), the UK body responsible for promoting the fishing industry, made available a pro forma document for fishermen to carry out the process themselves. However, this document attempts to cover the majority of risks on board fishing vessels and inevitably in doing so, includes many trivial risks. Consequently, a lack of emphasis is placed on the major risks when taken in the same context. This has the effect of fishermen trivialising the entire document and disregarding it.

In addition to this, the Maritime and Coastguard Agency, through the Sea Fish Industry Authority, has introduced an additional 1-day training course specifically designed to educate fishermen in the values of risk assessment. However, this course, at the moment, is voluntary, and the current take-up is not good.

There is also a need to audit to ensure the procedure is being carried out correctly. However, the problem with auditing, or even giving on board advice, is that the regulator may be viewed as having endorsed the assessment, and if an accident does happen, it will be no surprise where fingers will be pointed!

Noncompliance

Many fishermen in the UK are using consultants, for a fee, to complete the process (risk assessment) on their behalf and document it to comply with the regulations. However, once this process has been complete, rarely, if ever, do fishermen refer to the documentation again, defeating the whole object of a continued risk-based approach to safety.

Worse still, a good deal of fishermen (skippers, owners, and operators) have not completed or been involved in any kind of risk assessment. This is a poor reflection on a regulation that has now been in force for more than 5 years.

The main reason for noncompliance is that fishermen do not fully understand the process of risk assessment. They are unaware that one needs to be complete. They see little benefit in completing one; it does not add the catching capability of the vessel, and the documentation produced in order to help fishermen in its process is not user friendly. As long as this situation remains, risk assessments will have little effect on the improvement of safety in the UK fishing industry.

Enforcement

For many years in the UK fishing industry, a tenuous relationship has existed between the industry and the agency responsible for enforcement of regulations (UK Maritime and Coastguard Agency 2002).

One problem that appears to manifest itself in the fishing industry, probably more than in any other industry, is the difficulty in reaching agreement between the regulator and the industry itself. Even when agreement is reached, enforcement of regulations is not carried out. This inevitably leads to complacency both in the industry and in those responsible for enforcement.

In addition to this, the fishing industry has a very strong political lobby. Representatives of fishing communities are continually arguing in support of sustained economic well being, which masks safety to a certain degree. As long as profit remains at the forefront of fishermen's lives, it could be argued that any change in attitude to safety is a long way off. As far as government policy on safety is concerned, it meets industry representatives every 6 months in an effort to reach agreement on a range of related issues. However, progress is slow.

Benefits of risk assessment

The benefits to safety when risk assessments are properly conducted and involve persons familiar with a vessel's operation are considerable. They tend to be of little use when they are seen as a bureaucratic requirement, are conducted by outsiders, and sit on a shelf gathering dust after completion.

Risk assessments identify the known hazards, their potential consequences, the risk of such hazards occurring, and the control or safety measures to reduce the risks so far as reasonably practical. What is as far as reasonably practical is a decision that has to be made based on the risk involved balanced against the cost (in time, money, or inconvenience) to reduce the risk further. Ideally, this information should be passed to everyone concerned, thereby with safety awareness raised and risk reduced, a safer working environment results.

Risk assessment is a self-perpetuating process as lessons learned in the activity of commercial fishing are translated into the continued process to improve safety. When used proactively before the start of a new operation, the identification of appropriate safety measures through risk assessment can prevent learning by bitter experience.

Conclusions

The results of this research, which included several visits to fishing vessels that had been involved in accidents and exhaustive interviews with those skippers and crews, has shown that the risk assessment approach to safety in the UK fishing industry is not yet working.

The UK regulators and safety organisations are concerned about this fact and currently are working toward a different approach.

It is not envisaged that the requirement for a risk assessment to be conducted on all vessels is going to disappear. In fact, in an increasingly litigious society where personal injury claims are almost forming an industry in their own right, it is likely that the pressure for employers to minimise risk in the workplace will continue to increase.

Risk is assessed each day by many conscientious operators when they walk on board their vessels, identifying the hazards–be they related to weather or the condition of the vessel–and deciding what can be safely achieved and what safety precautions need to be taken. This is not a new process. The vessel operators are professionals more than capable of undertaking what is required.

Safety assessment is an imposition like any other prescribed regulation. However, a good assessment requires knowledge, motivation, perception of need, and a will to do it. It also requires a positive safety culture.

The challenges are to make fishermen aware of the requirement to conduct risk assessment and to produce guidance in a user-friendly format, encouraging fishermen to use it. As with the ISM Code, applicable to merchant vessels, there is also a need to audit. Only then can risk assessment start to have a positive impact on the safety of fishing vessels. Other avenues of getting the message across will probably need to be investigated.

Unfortunately, although I regret to say so, enforcement also has its part to play. If only one fishing vessel operator were prosecuted for noncompliance with the regulation, and this fact was publicised in the fishing press, it would probably have as much effect as all the education and training put together. However, it is appreciated that obtaining such a conviction when risk assessments need not be written would be difficult.

The fishing community must change its attitudes and adopt a positive safety culture. Every individual fisherman must be concerned about safety so that it becomes second nature to carry out basic checks and to correct things that are wrong. They must be seamen as well as fishermen, and they must resist the temptation to condemn anyone who suggests there are better ways of doing things.

With a change in practice in the way things are done it is hoped that the benefits of risk assessment will fall out of good practice. It is unlikely that risk assessment as it stands in the UK fishing industry today will lead to that practice unless a different approach is adopted.

Finally, it is in everyone's interest that all fishing vessels operate as safely as possible. Risk assessment can contribute toward this goal. It is a useful tool and of benefit to all.

Reference

Maritime and Coast Guard Agency (2002). Annual report and accounts, 2001-2002. House of Commons, London Stationary Office, United Kingdom.

Session Six

SESSION SIX:
SOCIAL FACTORS CONTRIBUTING TO WORK-RELATED INJURIES TO COMMERCIAL FISHERMEN

Vessel off the UK coast *(Photo courtesy of Cliff Brand)*

FATIGUE AND THE COMMERCIAL FISHING INDUSTRY: AN INTERNATIONAL PERSPECTIVE

Angela Baker, PhD
Sally Ferguson, PhD
Centre for Sleep Research, University of South Australia
Basil Hetzel Institute, Queen Elizabeth Hospital
Woodville Road
Woodville, SA, Australia

Background

Commercial fishing is considered to be one of the world's most dangerous occupations. From a global perspective the International Labour Organisation (ILO) in Geneva estimates that approximately 24,000 deaths and 24 million nonfatal injuries occur worldwide each year at sea (NIOSH 2000). In Australian waters, the work-related death rate in the commercial fishing industry is 89 deaths per 100,000 workers. This figure is 16 times higher than the all-industry rate for Australia of 5.5 deaths per 100,000 workers per year (Batchelor and Bugeja 2003). However, self-employed workers were not included in the analysis, suggesting that the number may be even higher. The UN Food and Agriculture Organisation (FAO) estimated the fatality rate in the fishing industry in Australia to be 143 per 100,000 (Australian Transport Safety Bureau 2001). In the United States of America, the fatality rate for commercial fishermen in 1998 was 151 per 100,000. The most recent analysis done in New Zealand examined fatalities between 1985-2000 and showed that the commercial fishing fatality rate was 167 per 100,000 (FISH Group 2003). Therefore, in relatively affluent and advanced countries such as Australia, New Zealand, and the USA, commercial fishing remains a highly dangerous occupation.

Great effort has been spent determining the causes and contributing factors to commercial fishing fatalities around the world. Issues such as environment, design and construction of equipment, and safety awareness are being addressed at various levels within the industry. However, another area where improvements can potentially be made is human factors. Human error has been attributed to 50% of all commercial vessel casualties and 70% to 80% of all marine accidents. Such errors include improper procedures, inexperience, poor judgment, carelessness, and navigational error. Contributing factors have been cited as stress, boredom, and fatigue (National Research Council 1991). Recently the ILO identified fatigue as a specific priority area to aid in the promotion of safety and health for those working within the fishing industry. Issues surrounding fatigue include long hours of work, few hours of sleep, physical and mental demands of the job, working at night, etc. A hazard occupational data sheet has been developed to address the issue of fatigue (Wagner 2000). Furthermore, the UN agency charged with the task of coordinating international maritime transport, the International Maritime Organisation (IMO), has introduced two conventions addressing fatigue. They are the Standards of Certification and Training of Watchkeepers, 1978 (as amended in 1995), and the International Safety Management Code (ISM Code).

What is fatigue?

Functionally, fatigue arises as a result of inadequate restorative sleep. Work-related fatigue is largely a consequence of hours of work that reduce the opportunity for recovery sleep and is generally worse when work hours are long or are undertaken at night. The human biological timing system is programmed for sleep during the dark hours and for activity during the daylight hours. For those individuals working outside the traditional hours of 0900 to 1700, Monday to Friday, achieving an adequate amount of sleep is difficult because of the time at which sleep opportunities occur. Research into shiftworkers' sleep, using both subjective measures (sleep diaries and questionnaires) and objective measures (activity monitors and sleep recording equipment), indicates that shiftworkers get significantly less sleep depending on the time of the particular shift (Siebenaler and McGovern 1991; Smith-Coggins, Rosekind et al. 1994; Smith-Coggins, Rosekind et al. 1997). In studies comparing sleep loss with alcohol intoxication, it was found that after 17 hours of wakefulness, performance decreased to a level equivalent to a blood alcohol concentration (BAC) of 0.05% (the driving legal limit in Australia and in many places around the world) and after 24 hours, performance decreased to a BAC of 0.1% (twice the legal limit) (Dawson and Reid 1997; Lamond and Dawson 1999).

Fatigue has been identified as a major root cause of human error in a number of extremely high-profile disasters. The Chernobyl nuclear meltdown disaster occurred at approximately 0120 (International Atomic Energy Agency

1986) and the Three-Mile Island meltdown at 0400 (Office of Inspection and Enforcement 1979). Both incidents resulted from human error. Workers did not respond correctly to either unexpected or unusual mechanical or control room failure, despite the circumstances being controllable. Furthermore, key management personnel achieved a minimal amount of sleep as a result of their previous work hours prior to the 1986 explosion of the space shuttle Challenger (Presidential Commission on the Space Shuttle Challenger Accident 1986). The Exxon Valdez grounding in Prince William Sound occurred because a crewman who was scheduled to be off shift at the time failed to recognise and act on a signal to turn the ship (National Transportation Safety Board 1990). Each of these accidents was largely contributed to by human fatigue. Such high-profile incidents have resulted in a greater awareness about the association of fatigue and human performance, particularly with reference to health and safety.

Reports indicate that fishermen routinely work 24 to 96 hours or more with little or no sleep (Committee on Fishing Vessel Safety 1991). Hence, it is not surprising that fatigue is a major problem within the industry. Fatigue has been shown to be responsible for at least 2% of human-related collisions such as that between the Handymariner (a bulk cargo vessel) and the Lipari (a fishing vessel) off the Western Australia coast on the 18 January 2001 and 4% of groundings (Australian Transport Safety Bureau 2001). Furthermore, six groundings per year in UK waters such as that of the cargo ship Jambo are believed to be fatigue-related (Marine Accident Investigation Branch 2003).

In the past, traditional approaches to managing fatigue in the workplace have focused on prescriptive measures (e.g., regulating hours, breaks, and time off duty). However, much of the research investigating fatigue in the workplace has found that from an operational perspective, these measures can be functionally problematic and restrictive and in some cases can actually make the problem worse. These prescriptive regulations are not based on the scientific understanding of fatigue. A more recent approach adopted in Australia and elsewhere is to move away from this traditional view toward one where organisations focus on performance-based outcomes in managing fatigue. For example, the Western Australian and Northern Territory governments regulate their commercial trucking industry in this fashion. Furthermore, the railways in the UK utilize a similar technique known as the "safety case," which provides flexibility and requires an auditable system to be in place. This outcomes approach recognizes the needs of different organisations and companies in different environments with different operational requirements and exposures. Hence, the emphasis is on reducing fatigue-related risk rather than solely focusing on hours of work. This shift has allowed managers to use the flexibility of the health and safety environment instead of bureaucratic rule-based systems to minimise and mitigate work-related fatigue without compromising economic productivity.

The human element in safety is the responsibility not only of the employer but also of the individual. The fishing industry and those tasked with the responsibility of improving safety standards face a somewhat unique challenge among industries. The ILO (1999) states that—

> Many fishermen have a different perception of danger [than] shoreside workers. Social and cultural attitudes, beliefs and values play an important role in the perception of, and response to, danger. The denial of danger, independence, fatalism, the belief that safety is a problem that primarily requires a technological solution, are common themes among fishermen. Efforts to improve safety should begin with trying to understand the fishing culture and to involve the fishers in the development and enforcement of safety regulations.

Thus, an international project aims to investigate the role, effect, and consequences of fatigue and long hours of work on the health and safety of those working within the commercial fishing industry in Australian and international waters. Initial work for this project has commenced in determining team membership, background information on the industry, and possible methodologies to be used. It is envisaged that the lead-in time for this international work would be no less than 12 months from August 2003. This lead-in time will be spent viewing simulators, planning scenarios, and most importantly, developing collaborative relationships with a variety of industry stakeholders such that the research is targeted and focused.

References

Australian Transport Safety Bureau (2001). Independent investigation into the collision between the Hong Kong flag bulk cargo vessel *Handymariner* and the fishing vessel *Lipari* off the south coast of Western Australia on 18 January 2001.

Batchelor M and Bugeja L (2003). Commercial vessel fatalities in Victoria 1991-2001. Victoria, Australia: State Coroner's Office.

Dawson D and Reid K (1997). Fatigue and alcohol intoxification have similar effects on performance. *Nature* 38: 235.

Fishing Industry Safety and Health Advisory Group (2003). Final Report. MS Authority, New Zealand.

International Atomic Energy Agency (1986). Summary report on the post-accident review meeting on the Chernobyl accident (25-29 August 1986, Vienna, Austria).

International Labour Organization (1999). Safety and health in the fishing industry: Report for discussion at the Tripartite Meeting on Safety and Health in the Fishing Industry (Geneva, Switzerland, 13-17 December 1999). Geneva: International Labour Office.

Lamond N and Dawson D (1999). Quantifying the performance impairment associated with fatigue. *Journal of Sleep Research* 8: 255-262.

Marine Accident Investigation Branch, United Kingdom (2003). Report on the investigation of the grounding and loss of the Cypriot-registered general cargo ship *Jambo* off Summer Islands, west coast of Scotland 29 June 2003.

National Institute for Occupational Safety and Health (2002). Public Health Summary. *In* Proceedings of the International Fishing Industry Safety and Health Conference (Woods Hole, Massachusetts, Oct. 23-25, 2000), Lincoln JM, Hudson DS et al., eds. Cincinnati, OH: National Institute for Occupational Safety and Health. DHHS (NIOSH) Pub. No. 2002-147.

National Research Council (1991). Fishing vessel safety: Blueprint for a national program. Washington DC: National Academy Press.

National Transportation Safety Board (1990). Grounding of the US tankship *Exxon Valdez* on Bligh Reef, Prince William Sound, near Valdez, Alaska, March 24, 1989. Washington, DC: Government Printing Office

Nuclear Regulatory Commission, Office of Inspection and Enforcement (1979). Investigation into the March 28, 1979, Three-Mile Island accident by the Office of Inspection and Enforcement. Washington, DC: Government Printing Office

Presidential Commission on the Space Shuttle Challenger Accident (1986). Report of the Presidential Commission on the Space Shuttle Challenger Accident. Washington, DC: US Government Printing Office.

Siebenaler MJ and McGovern PM (1991). Shiftwork: Consequences and considerations. *AAOHN J* 39(12): 558-67.

Smith-Coggins R, Rosekind MR et al. (1997). Rotating shiftwork schedules: can we enhance physician adaptation to night shifts? *Annals of Emergency Medicine* 4: 951-961.

Smith-Coggins R, Rosekind MR et al. (1994). Relationship of day versus night sleep to physician performance and mood. *Annals of Emergency Medicine* 24: 928-934.

Wagner B (2002). Safety and health in the fishing industry: An ILO perspective. *In* Proceedings of the International Fishing Industry Safety and Health Conference (Woods Hole, Massachusetts, Oct. 23-25, 2000), Lincoln JM, Hudson DS et al., eds. Cincinnati, OH: National Institute for Occupational Safety and Health. DHHS (NIOSH) Pub. No. 2002-147.

FAMILY PREPAREDNESS PREVENTS
ACCIDENTS AT SEA

Anne Melton
USCG Auxiliary
Community Emergency Response Team (CERT)
Bellingham, Washington, USA

Worrying about family and home problems while fishing can cause stress and inattentiveness to one's work and surroundings, jeopardizing the safety of a fishing crew. Concern for those at home is no longer brought on only by such crises resulting from bills that need to be paid, cars breaking down, or a leaking roof. What can be done to decrease the added mental burden brought on by the worry of intended attacks or a major natural disaster affecting the family back home?

Preparedness is the first line of defense against accidents when fishing and should hold the same importance for crises at home. Everyone should be prepared for at least 72 hours of no assistance from emergency responders. An earthquake, for example, may result in bridges being down, phones inoperative, and broken water mains. If there is no electricity, gas cannot be pumped for cars; grocery stores cannot operate their computers to sell food; heat may not work; and there may be no way to cook food. In some disasters, entire homes may be lost. If there is infrastructure collapse and no chance of getting help for several days, each family will be on its own for support and care. Would your family know what to do if you were not there to help?

Start now to prepare for the event that you hope never happens. Because those who are fishing will want to contact their loved ones as soon as possible, an out-of-area contact will be the first item. During a disaster, each member of the family can call the selected out-of-area person with information to pass along to each other. That will allow everyone to know the whereabouts and condition of each other.

Collect essential items and place them in a transportable kit: battery-powered radio, water, nonperishable ready-to-eat food, flashlights, extra batteries, medicines, first-aid supplies, a change of clothes and shoes, blankets or sleeping bags, and games or books. Have toilet tissue, wet cleaning wipes, and additional water to be used for hygiene. Can openers and eating utensils are needed. Don't forget the special baby items and pet's food, carrier, and leash. Individual preferences and needs should be considered and included. Plastic bags of various sizes are good for many purposes: small ones keep organized medicines, contact lenses, and hearing aid batteries. Large garbage bags can serve as a windbreaker, rain coat, or even as additional warmth over a sleeping bag. Large plastic bags can hold contaminated material and waste for later disposal or burial. When preparing an emergency kit, you should make it light enough to carry in case the roads are impassable to vehicles and the home must be abandoned. A backpack for each person would distribute the weight and provide participation and understanding of the preparation for all family members.

Every family should be prepared to evacuate at a moment's notice. Reasons to evacuate could include a fast-approaching forest fire, a hostage stand-off, accident or derailment of a train transporting dangerous materials, flash flood, and a home that is otherwise damaged and too dangerous to occupy. When the emergency supplies are already collected, the family members can leave with enough provisions for the first days. They are prepared to have the basics as they wait out the catastrophe.

In addition to preparing for evacuation, a family should also prepare to "shelter in place." A chemical spill (accidental or intentional), for example, may lead to an emergency announcement to stay in until the chemicals dissipate. A shelter-in-place kit includes materials to seal one room from outside air—close off heat or cooling vents, duct tape plastic sheeting around windows, place a towel at bottom of doors, duct tape around doors, tape over electric wall outlets. Keep the radio on (battery power available) and flashlight handy in case electricity is lost. If an announcement is made to evacuate, the family must leave immediately with their emergency supplies and go only on the route designated by the emergency responders.

After a disaster, emergency responders and relief workers will be on the scene, but they cannot reach everyone immediately. If the family is isolated or confined due to natural disaster, infrastructure collapse, or because emergency responders are overwhelmed, they must be prepared to cope with emergencies until help arrives. When the family is apart for periods of time and disasters occur, there may be less stress if emergency preparations are ready and everyone has a plan in place to respond. Make a kit, make a plan, and stay informed. Safety at home is necessary to have safety at sea.

Additional information is found on these Web sites:
http://www.fema.gov/areyouready/
http://www.redcross.org/services/disaster/beprepared/hsas.html
http://www.ready.gov
http://training.fema.gov/EMIWeb/CERT/certfog.asp

A DESCRIPTION OF LIFE STYLE FACTORS AFFECTING CREWS WORKING ON DISTANT WATER TRAWLERS

Anna Maria Simonsen, RN
Faroe Islands

Background

Commercial fishing and offshore fish processing are high-risk work. The risk of accidents and sickness in the offshore fishing industry is up to ten times higher than "land-based" high-risk occupations. Fishermen's work differs greatly from "normal" working life on land. Fishermen find themselves, often for months on end, in a limited space aboard their ships without the possibility of going ashore. They are exposed to unstable workplaces with constant noise and vibration. In circumpolar areas, workers are exposed to nearly continual light or darkness throughout their shifts, depending on their vessel's location and the season of the year.

Commercial fishermen work, eat, and sleep on board their vessels. There are no free days or weekends, and if there were, crew would have to enjoy them at their workplace. There are limited possibilities for exercise by crew, both because of too little free time and too little space, and because (taking into consideration the circumstances) motivation is not particularly high. Social interaction can be forced; you have to live together with your colleagues, whether you like them or not. Crew members are faced with physically, psychologically, and socially straining, continuous shift work, which often carries with it elevated risks for injury.

Another factor that can have a negative effect on health is the food eaten by the crew. If mealtimes are one of the few highlights in an otherwise busy and monotonous day, they may become compensation for the stresses experienced by the crew. "Good food" is easily equated with too much, often unhealthy, food containing high amounts of animal fat. Some crew members indulge in excessive use of stimulants such as coffee and tobacco, both of which carry consequences to health. Commercial fishing also entails loneliness and deprivation, as there are no opportunities for physical contact with families and participation in the social and cultural life at home.

For crew members, coming home can easily comprise a mixture of emotions, both good and bad. Without doubt, commercial fishers are happy to be with their families and friends again and to take part in daily life. But their biological clocks must be readjusted to a normal daily rhythm after months of shift work, and this can be hard for many who have become used to the ship's constant movements, vibrations, and noise as well as always hearing the ship's movements, even in sleep. Many crew members end up with sleep problems.

Being home again also means a big upheaval for family life. Spouses must adjust to each other's presence and learn how to share responsibilities and tasks, such as bringing up children, economy, maintenance of property, etc. Many crew members choose to stay at home for one trip if there is a large family event ahead, such as a wedding, birthday, confirmation, etc. But this choice has a cost in lower annual income.

The Faroese fishing fleet contains ships of various sizes and functions. The way in which they work, in which distant waters they fish, and the length of the fishing trips varies greatly. Some of the fleet fish close to the Faroes and land fish about every 14 days. The crew on these ships must, if the fishing is good, work for several days without sleep. However, the subjects of the study described in this paper all work aboard long-distance trawlers, where one fishing trip can last from 2 to 4 months, traveling far from the Faroes, without any port calls. In 2000, there were five such trawlers working from the Faroes, producing salt fish and fillets. The hypothesis of the study described here was that Faroese fishermen aboard long-distance trawlers would be subject to a number of lifestyle factors that could have implications for their health status.

Methods

To more fully understand factors associated with the health of crew members on long-distance trawlers, we selected a random sample of 35 crew, working aboard all five Faroese long-distance trawlers, whom we asked to participate in a survey. The project's investigator developed a questionnaire aimed at examining the lifestyle factors experienced by crew members of a Faroese long-distance trawler. The questionairre contained items about exercise, food intake, cigarette and caffeine consumption, leisure time activities, and attitudes toward work. All written communications concerning this project were in Faroese.

The format of the questionnaire was designed to identify the presence of life style factors that I hypothesized would have negative impacts on fishermen's health. I obtained the permission of the ship's manager and captain to start the project before I sent the questionnaire out. An informative letter was sent to the captain and the crew. As part of a review process, the Faroese Fishermen's, the Engineers', and the Master Mariners and Navigator's unions were also informed in writing. The Faroese Science Ethics Board was informed about the project.

Before the questionnaire was sent out, a letter with information on what the project was about was sent on board each of the five trawlers. The questionnaire was sent by fax with information about when the answers should be returned. Of the 35 men who were asked to answer the questionnaire, 29 chose to take part. The age of respondents was fairly young: 3.5% were under 25, 21% were between 25 and 29, 48% were between 35 and 44, and 27.5% were 45 years or older.

Results

Exercise
Survey questions asked about respondents' exercise patterns while aboard long-distance trawlers. Only one respondent stated that he exercised regularly, working out about 120 minutes a week. The same person had a normal body mass index (BMI). The vast majority of respondents, 28 out of the 29, did not exercise regularly.

Food intake
We asked respondents to provide data on weight and height. Research investigators later converted this data into age-adjusted BMIs for each respondent. Based on analysis of respondents' data for weight and height, our study found that about 57% had BMIs that were higher than normal weight ranges.

Cigarette and/or caffeine intake
Respondents were asked if they smoked, and if so, how many cigarettes a day. All respondents were asked if they shared a room with a smoker. If so,

this study presumed such respondents to be passive smokers. Another survey item asked respondents to provide information on the number of cups of coffee each one consumed during the day.

Our survey found that there were fewer smokers on board long-distance trawlers and that nonsmokers comprised the majority of respondents. There were 48.3% smokers and 51.7% nonsmokers. Of the smokers, 71% smoked 20 or more cigarettes a day. Our findings were compared with those of an extensive annual questionnaire of all 9th grades in the Faroes carried out since 1989 by the Department of Occupational and Public Health in collaboration with the Faroese School Health Nurses. In this survey, respondents stated that 42% of their fathers smoked. If this statistic is reflective of Faroese men as a whole, it appears that fishermen on Faroese trawlers smoke more than Faroese men on average. This is interesting, when considered against the production tasks that take place aboard long-distance trawlers, where smoking bans are in place at production lines and where working hours limit cigarette consumption.

Coffee consumption among respondents varied, from none to up to 10 mugs per watch.

Shift work

Respondents were asked which watches they worked, how often the watches were broken, and whether they judged their work to be monotonous or not. We asked respondents to tell us the number of days they worked in the past year. The two watches were evenly represented in the answers received: 15 worked on the master's watch, 13 on the mate's watch, 1 chose not to answer this question.

The question on how often the watch was broken up during the week was incorrectly formulated in the questionnaire, and a third of the answers received were marked with a question mark. Therefore, it is not possible to say how often broken watches occurred.

The majority of respondents (23) did find that their work was monotonous, five said no, and one did not respond. Working hours aboard trawlers varied considerably among the respondents.

Isolation

Respondents were asked to identify an acceptable period of time for duty aboard long-distance trawlers and whether they wanted to continue to work as commercial fishermen. We also asked respondents to describe the ways in which they passed the time during long voyages. Feedback from respondents about optimal length of time aboard long-distance trawlers indicated that any period above 4 months would be considered "too long." Most respondents (21, or 72.4%) felt that 2 months was an acceptable length for an individual voyage.

Of the 29 respondents who replied to the survey question asking about career choices, 25 wanted to remain at sea and only four were planning to stop (86.2% and 13.7% respectively). This may be because Faroese commercial fishermen like this lifestyle and the freedom and excitement involved in commercial fishing. But it may also be a matter of the limited earning possibilities in other occupations. While fishermen's earnings can be high, the number of hours that they work must also be compared against land-based occupations.

Respondents' leisure time pursuits, in order of selection, were TV, reading, different types of games, other, and exercise. The types of television programs watched varied, with movies frequently shown. When trawlers receive mail deliveries from other ships, there were also video recordings of news from home.

Conclusions

The results of this study indicate that my hypothesis that trawler crews experience a number of negative life style factors is well founded.

1. The vast majority of respondents did not exercise.
2. Most of the respondents were over normal weight limits, as documented by BMIs.
3. It appears that there are more smokers among the respondents then would be expected.
4. The number of hours worked per year by respondents is far greater than those worked in land-based occupation.
5. The difficulties associated with the lifestyles of commercial fishermen aboard long-distance trawlers may contribute to early retirement from

this occupation, as evidenced by the age distribution of the respondents, which was skewed to younger ages.

There are many ways to promote healthy work environments for commercial fishermen aboard long-distance trawlers. The findings of this study can be likened to a "buffet" of recommendations for a future worksite health and safety plan for this industry.

Prioritizing is important in order to obtain improvements. The first priority, and one that could definitely improve conditions, would be to limit the length of any one voyage on board long-distance trawlers to 2 months. The majority of the crew wishes this, and without doubt this is also the wish of the individuals' families. We are pleased to note that the Faroese Fishermen's Union has called for a 2-month limit, as well.

A second priority should be made to promote both healthy diets and exercise. Ships' managers and cooks should receive professional help from nutritional experts. It should be possible to provision a ship so that products with high animal fat content are limited. Educational programs for crew, with information and guidance on diet and exercise, could stimulate the motivation for crew members to make more nutritional food choices and to implement exercise programs.

We hope that the findings from this project contribute to the improvement of the work environment for trawler crews. There clearly is room for improvements to promote healthier life styles for workers aboard long-distance trawlers. We look forward to additional research projects with this population, and expect that, with legislative support, many changes can be made to enhance the lifestyle experienced aboard these vessels.

The Occupational Medical Department of the Faroese Hospital Service subsidized this study.

Appendix: Changes in lifestyle factors of Faroese fishermen

Fishing has been the main industry of the Faroes for hundreds of years. However, vessel design and amenities for crew members have changed dramatically. Numerous differences exist between the old fishing boat, where work was carried out on open decks in all types of weather and where the crew slept crowded together in small cabins in which they made food, dried clothes, talked, and smoked to the super-modern factory trawlers of today, where working conditions have improved, the crew lives in one- or two-man cabins, the galley and the recreation rooms have everything a crew member could wish for, and living conditions include washing machines and tumbler dryers, and, on some ships, a sauna and exercise room.

Life for the Faroese fisherman has generally been typified by hard physical work, long work days, little sleep, and at times monotony, bad diet, loneliness, and deprivation. There have always been considerable risks of attrition, sickness, disablement, and shipwreck. As the main part of Faroese industry is and has been fishing, the choice of this occupation may be the result of social and/or familial pressure, but it must also be pointed out that the Faroese have always been, and still are, an extremely seaworthy people. Many times they have been pioneers testing out new techniques in commercial fishing, for example, as crew and skippers when shrimp fishing with trawls began in Greenland. To clearly delineate the changes in working conditions, we conducted interviews with two participants. The first is a retired commercial fisherman who worked on board Faroese commercial fishing vessels in the 1950s and 1960s; the second is a young man who is currently working as a commercial fisherman.

Interview with a former fisherman

"We worked 12 hours at a time, 6 hours free time. The free time included two meal times, we never slept more than a good 5 hours.

"For a short time we tried a watch system called the Seawatch. Some of our colleagues who had sailed with cargo boats wanted to try this. We were divided into two teams. Team A had the dogwatch, from 2400 hours to 0400 hours. Team B had the watch called the Seawatch, from 0400 hours to 1200 hours, then team A was on watch again from 1200 to 2400 hours. This watch system was good for the crew. The drawback of this was that both teams ate almost at the same time, so if the fishing was good, there was no work on the deck at that time. Therefore we could not keep this watch system. The watch system we know now with 6 hours on and 6 hours off began around 1956.

"We had newly slaughtered meat, both sheep and beef at the beginning of a fishing trip, which could have been 4 to 6 months long. When this was eaten up we had salted meat with pea soup and then tinned beef, corned beef. We also had bacon. We ate a lot of fish as well as birds.

"Salted whale meat and blubber were also a part of the provisions bought by the ship's managers, from men who were good at catching whales and, according to how big the whale catch was, were able in this way to earn money. We got potatoes but they were often of doubtful quality and often didn't stretch the whole trip. We did not have any other fresh vegetables, only dried peas and herbs. We only had raisins, prunes and the occasional dried apples for fruit. Bread and cakes were baked on board. We only got milk when we could buy dried milk. Cream was the tinned cream called Molly. Some of us had dried sheep's meat from home. Some ship's managers bought dried sheep's meat for the crew.

"Boots and outer garments were as they should be, we also had good woollen clothes. However, gloves were a big problem. They were attacked by water fleas (Cyclops), which made them water absorbent, resulting in destroyed, raw hands. Our hands were often painful, because we had to continue working anyway.

"We got news from home when we went into a port to bunker, or from other ships from home. We occasionally received telegrams, which could be a difficult process. Not every home had a telephone then so the option to call home was limited.

"We took a book chest from the library with us; several of us read a lot. We also played cards and chess during off duty. Hobbies such as making small dolls' houses, cradles, model ships, etc. were very popular. Sunday should be a day of rest. We observed this by holding a service at the time when the least amount of work was done. Otherwise, work was carried out like any other day."

Interview with a young man currently working as a fisherman

"I have chosen this career because I have grown up with it and it is in my blood. Even though it is hard at times, it is exciting—how much we fish, what price we will get when the catch is sold, whether we fish more than other ships and so on. In addition you can earn a lot of money.

"When we begin a new trip the crew is divided into two watches that are called the master's watch and the mate's watch. There are no fixed rules about how we are divided, it depends upon splitting up experience and skill as good as possible. The master's watch works from 0600 to 1200 hours and from 1800 to 2400 hours. The mate's watch works from 1200 to 1800 hours and from 2400 to 0600 hours. This is how the whole trip is run; you do not change shifts. Sometimes when it is very busy, work is carried over so that one half of the watch works 3 hours more and the other half of the watch starts 3 hours earlier, so you only get 3 hours off duty.

"You quickly get into the watch rhythm, you soon get used to it. The problem is when you come home again and try to live a normal daily rhythm again. It can be difficult. The main meal times on board are the same as at home, lunch is eaten at 1200 hours and supper at 1800 hours.

"Working on board is not without danger- you should watch out all the time, as a serious accident could easily happen. Working on the trawl deck is dangerous—you should always be on the alert and aware of what you are doing. A main rule or law, you can say, is to look out for yourself and your safety, as no one else will do it for you.

"Rooms are intended for two people, although some times every man gets a room for himself. However the master, mate, chief engineer, cook and telegraphist always get their own rooms. We change bedding twice a month and we keep our rooms clean. We have a washing machine and tumbler dryer, and we wash our own clothes. We take turns in cleaning- we clean the passageway floors, and recreation room. There are common bathrooms on the passageway. I think on one of the newer ships they have a sauna and an exercise room.

"The mood on board can swing according to how the fishing is going and how long the time away from home. It also depends upon whether one hears good or bad news from home. Sometimes you can have an argument but it is nearly always forgotten by the next watch. The food on board is good enough, you don't go short of anything. Some drink a lot of coffee, others not at all.

"Smoking varies; there are several who do not smoke. Smoking is forbidden on the production line and also there are some smoke-free areas on the ship, smoking is also forbidden in the recreation room. These prohibitions are strictly enforced.

"There is a service on Sunday, varying a little on who reads. Once we had a priest with us, then of course he read. We also sing a couple of psalms when we hold a service.

"We see some television, we also have video films from home. There is a book chest on board, some have their own books and papers with them. We also play cards and chess, also between different ships. We can buy work clothes on board at cost price.

"We receive mail on ships that have just been home, it may be once or twice a trip. So we get letters, papers and parcels, we also get Christmas gifts in this way sometimes."

SAFETY ISSUES IN ARTISANAL AND SMALL-SCALE COASTAL FISHERIES: A CASE STUDY FROM THE BAY OF BENGAL REGION

Dr. Yugraj Singh Yadava
Director
Bay of Bengal Programme Inter-Governmental Organisation
91, Saint Mary's Road, Abhiramapuram
Chennai – 600 018
Tamil Nadu, India
E-mail: bobpysy@md2.vsnl.net.in

The Bay of Bengal region

The Bay of Bengal, located in the monsoon belt, is bounded by eight countries: the Maldives, Sri Lanka, India, Bangladesh, Myanmar, Thailand, Malaysia, and Indonesia. About one-quarter of the world's population resides in the littoral countries of the Bay of Bengal, approximately 400 million of whom live in the bay's catchment area, many subsisting at or below the poverty level. An average of 65% of the region's urban population lives in large coastal cities, and migration toward the coastal regions is increasing.

The Bay of Bengal, including the Andaman Sea and the Malacca Straits, covers an area of 2.83 million sq km. The total length of the coastline of the countries straddling the bay and its adjacent seas is about 200,000 km. The coastal and offshore waters of the region support numerous fisheries, which are of great socio-economic importance to the countries bordering the Bay of Bengal. Among the most important of these are the inshore small pelagics, demersal, and shrimp fisheries and the offshore fisheries for tuna. During the last decade, fish landings from the region has increased by over 60%, with the latest statistics indicating catches exceeding 3.7 million tonnes. More than 300 fish species are estimated to be of commercial value.

Small-scale fisheries in the Bay of Bengal are still largely traditional. Log rafts and wooden crafts are popular in coastal India, Bangladesh, and Sri

Lanka. Evolved over centuries, these artisanal, low-cost, environmentally friendly fishing crafts are next to perfect. Most of the fishery resources of the coastal waters in the bay are heavily exploited due to unregulated fishing activity and open access. This is further complicated by the growing population, poor national resource management strategies, and a general lack of knowledge and data on the functioning of the ecosystem as a whole.

The Bay of Bengal, unlike many other seas, is rough for most parts of the year. Cyclones are frequent and come without much warning, and the monsoon winds increase the perils of fishing at sea. The artisanal and small-scale fishing vessels are unequipped to meet the challenges of the offshore waters of the bay. With the resource getting scarce in the coastal waters, the fishermen are venturing deeper into the sea, risking their lives. Some get drifted and end up in an alien land. Some fail to return, leading to a long and tortuous vigil for the family. There are few who survive the ordeal and return. Most perish and leave a widow and children destitute. The misery for the family unfolds immediately. With little savings and practically no alternative means of livelihood support and income opportunities, the family is either forced to migrate to the urban areas to work as labourers or beg for a living. The debt of the moneylender also makes many families work as bonded labourers.

Fishing-related deaths are on the increase and are more likely the result of economic pressures and human factors, such as risk-taking, fatigue, stress, or simply attitudinal problems. A high risk for loss of life has been accepted as part of the fishing culture in the region. Typically, small-scale fishermen are not registered; they are self-employed, and they have no safety net at all. The spirit of bravado has become a defence mechanism for most fishers.

Excessive fishing, lack of vessel maintenance, recklessness, and the gradual erosion of traditional skills have all led to fishing becoming the most dangerous occupation in the world (Turner and Petursdottir 2002). Accidents take place mostly because of engine failure, navigational difficulties, rudder damage, fuel shortage, capsizing, down-flooding (holes in the deck, for example), fire, explosion, collision, overloading, lack of safety equipment, etc. Many lives that are lost at sea could be saved if simple safety and communication equipment were kept onboard. The reasons for this human tragedy are all well known: lack of sea safety measures in the artisanal and small-scale fishing crafts.

Fishermen and fishing vessels are largely excluded from the vast majority of the provisions of the international shipping conventions drawn up by bodies like the International Maritime Organisation. Among commercial vessels and industrial fishing boats, a large work force and strong maritime unions bargain effectively for better safety and welfare measures. However, within the artisanal and small-scale fishing industry, workers are unorganised and at the mercy of the middlemen and the boat owners, who care little for their safety at sea. For those individuals or small groups of fishermen who own fishing vessels, the turnover and profits are so meagre that there is little left to use for safety and navigation equipment. The result: A risky adventure into the sea every day and thousand of lives lost every year.

Regulations governing boat construction, availability of on-board safety and navigation equipment, and timely warnings on rough weather are mostly lacking in the countries surrounding the bay or, if present, are not strictly enforced. Therefore, with the bad weather and the turbulence in the sea, either the boats capsize or they drift and end up in neighbouring countries. Apprehensions take place, and the fishermen languish in jails and detention centres for long periods since the prevailing legislation and penal codes governing maritime infringements in many countries of the Bay of Bengal region involve lengthy and cumbersome procedures.

Who is responsible for this? Government, the boat owners, or the fishermen? All three entities are equally responsible. The case of the artisanal and small-scale fisheries is perhaps the most pertinent in terms of promotion of responsible fisheries operations and the most problematic because safety regimes are the weakest here.

Keeping in view the fact that fishing is the most dangerous occupation in the world and that there is a significant and growing problem relating to sea safety, the Bay of Bengal Programme, (BOBP) (see note a) in association with the Fishery Industries Division of the Food aand Agriculture Organisation (FAO) in Rome, organised a regional workshop on Sea Safety for Artisanal and Small-sale Fishermen in Chennai during 8–12 October 2001. Forty-three participants from all the BOBP countries–Bangladesh, India, Indonesia, Malaysia, Maldives, Sri Lanka, and Thailand–as well as experts and resource persons, attended the 5-day programme. The significant output of the regional initiative was the Chennai Declaration on Sea Safety, which resolved to meet the challenges effectively through holistic fisheries management, adherence to mandatory requirements, installation of regulatory mechanisms, community involvement, and education and training.

Based on a review (see note b) conducted prior to the workshop and the discussions that took place therein, a summary on the existence, status, and effectiveness of formal and informal arrangements for ensuring the safety of artisanal and small-scale fishermen in the countries in the Bay of Bengal region (except Myanmar) is given in the following paragraphs.

Bangladesh

Overall, fisheries contribute 4.7% of the national GDP, 8% to the export earnings, and 60% of the animal protein intake in Bangladesh. These percentages represent a significantly higher contribution from the fisheries sector than in any other BOBP member country.

In the sixth 5-year plan (2000-2005), the objectives identified were to increase fish production, improve protein access, increase foreign exchange earnings and employment, and adopt a participatory approach to resource management. Resource management instruments currently in place include a ban on estuarine bag net and push net fishing, adoption of a closed season from January 15 to February 15, and a ceiling on the size of the shrimp trawler fleet. In 1982, Bangladesh signed the United Nations Convention on the Law of the Sea and in the next year promulgated the Marine Fisheries Ordinance for the management and conservation of marine fisheries resources. The rules are enforced by the Director of Marine Fisheries.

The total number of artisanal and small-scale fishing boats in Bangladesh is reported to be between 27,000 and 30,000. Of these, approximately, 10,000 are mechanised and 17,000 nonmechanised. Nonmechanical boats propelled by sail and oar range from 5 to 20 metres in length and are typically engaged in day-to-day fishing in estuarine waters. Set bag nets, stake nets, or beach seines are used. Mechanised boats vary from 5 to 15 metres in length and have engine power ranging from 22 to 60 bhp. These vessels are engaged in drift net, longline, or marine set bag net fishing, and remain at sea for 4 to 6 days.

The bulk of the marine catch is landed by traditional craft, either plank-built or dugout. Plank-built boats are mainly of two types—*dinghi* and *chandi*. The *chandi* is one of the most widely used traditional fishing boats in Bangladesh. It is a round bilge, carved planked, open boat with a high sheer aft. It is usually constructed by stapling shaped planks, after which the framing is nailed into position. The boats are decked with split bamboo, and a thatched bamboo shelter is affixed slightly forward of amidships.

It was observed that swamping occurred most frequently in steep head seas with the foredeck getting buried as opposed to being broached. No reports of losses attributable to fire were reported, although open wood fires are used for on-board cooking. None of the vessels carry fire extinguishers. A recent practice adopted as a low-cost approach to the mechanisation of small traditional boats has been to fit tube-well engines. These vessels are considered a hazard both to themselves and to the other boats operating in the crowded waters.

The other main cause of fatalities at sea is pirates. The Chittagong Port Authority reports in excess of 50 casualties each year attributable to piracy.

Two primary regulatory agencies having a role in the fisheries sector are the Marine Fisheries Office of the Department of Fisheries (DOF) and the Mercantile Marine Department. The DOF has been expanding with new offices being established in the six most important coastal fishing areas. Monitoring, control, and surveillance are currently limited to the activities of the DOF staff and the Marine Fisheries surveillance check posts. The DOF has no at-sea or air surveillance capacity, vessel identification requirement, or fully developed fisheries data system in support of an effective monitoring, control, and surveillance capability.

Bangladesh and India are the two Bay of Bengal countries subject to the most severe weather conditions associated with tropical cyclones. Sea conditions in Bangladesh are further aggravated by the seabed topography, which funnels tidal action into the shallow coastal regions, resulting in high levels of turbulence. Swell heights are reported to reach 5 meters with storm surges up to 7 metres. Bangladesh has built a strong forecasting and emergency response capability. Observatories are maintained throughout the country. Weather radar stations have been established in four locations. Satellite data are received from Japanese and Indian satellites and other international sources. The Meteorological Department maintains a 24-hour operation. Regional forecasts are issued twice daily, and marine forecasts, which provide wind speed and direction, with a qualitative description of the state of the sea, are issued daily. In the event of a cyclone, storm warnings are updated every 3 hours.

The significant risks in terms of sea safety in Bangladesh relate to over-exploited resources with declining returns to the fishing enterprise, the need to coordinate the regulatory environment, enforcement of safety standards, the need to look at alternate designs for vessels, upgrading communication equipment, piracy, and the development of a network of community organisations to address sea safety issues at the community level.

India

India has an open-access fisheries resource management regime. Under the Seventh Schedule, Article 246, of the Constitution of India, fisheries within the territorial waters of the country are under the jurisdiction of the coastal state government, and fishing beyond the territorial waters is under the jurisdiction of the union government.

The tenth 5-year plan (2002-2007) for the fisheries sector includes the objectives of optimising production and productivity, increasing export, generating employment, improving the socio-economic conditions of the fishermen and fish farmers, conserving aquatic resources, and increasing the per capita consumption of fish. A strong scientific capability and a network of fishery research institutes, which provide information on the marine environment, support resource management activity in India. The Department of Ocean Development is now focusing on areas such as satellite monitoring of sea surface temperature, identification of potential fishing zones, etc.

The total marine fish production in India has grown from 0.88 million tonnes in 1960 to 2.83 million tonnes in 2002-2003. The harvesting sector has evolved rapidly from a fleet of traditional canoes, dugouts, and small-planked boats to the development of a powerful fleet of small trawlers. The traditional Indian fishing fleets operating in the Bay of Bengal comprise about 128,400 traditional craft, of which about 16,300 are motorised. *Kattumarans*, which dominate the traditional crafts, are widely used on the east coast of India, from Puri in Orissa to Cape Comorin in Tamil Nadu on India's southern tip. The large number of *kattumarans* supports about half a million fisherfolk and accounts for a substantial quantity of the total marine fish production from the east coast of the country.

The *kattumaran* adapts itself to the harsh conditions of India's east coast, which features a high and dangerous surf and a lack of adequate landing facilities because of the constant movement of sand drifts. The *kattumaran* easily penetrates through the breakers, instead of riding over them, thus avoiding capsizing. It is, in fact, nearly unsinkable. It requires little upkeep since the hull does not have to be painted and caulked.

The *kattumaran* is the craft best suited for heavy surf beach conditions. Attempts to replace these traditional beach-landing craft with competitively priced surf-landing boats have so far been unsuccessful. Even in the very long term, the *kattumarans* are likely to be active in the near-shore fishery of the east coast states of India. Motorisation of *kattumarans* with outboard engines has helped the fishermen to overcome the drudgery of long trips, go faster and further into the sea, and navigate rough weather.

The crab claw sail is an old idea from the shores of Bay of Bengal. For a long time, experts have regarded this sail as primitive. But current wisdom regards the crab claw sails of *kattumarans* and outrigger canoes as very effective for navigation and propulsion. Besides the *kattumarans*, the popular traditional fishing crafts of the Bay of Bengal are *navas, masula* boats, *dinghies,* and *dhonies*.

The most remarkable traditional fishing craft anywhere are the *shoe-dhonis* of Kakinada in Andhra Pradesh. Shaped like shoes, they are both boats and homes. Each of these boats houses a fisherfolk family for almost 8 months a year. The *shoe-dhoni* families make a living by collecting shells from the Kakinada bay and selling them to traders. They return to their villages only for festivals and during floods.

From or to their villages, they journey through mangrove swamps for 20 to 30 hours, using sails and long poles. Women and children tend the rudder and even help in punting. The women also do the cooking, of course; none of the men can cook. Pregnant women continue life on the *dhoni* till the seventh month. Sometimes the children are born in the *dhoni*. These kids learn to swim before they can walk.

Monitoring, control, and surveillance responsibility in coastal waters is the responsibility of the adjacent state, with the Coast Guard having the responsibility for offshore waters. The Coast Guard operates a highly professional service from 15 bases around the coast. The country has also established an integrated capability for disaster management. This capability includes communication, decision-making, and the immediate mobilisation of resources by all the agencies at various levels of government. The Coast Guard is the lead agency for search-and-rescue operations relating to artisanal and small-scale fishers. Ninety percent of the distress calls are due to mechanical breakdown or running out of fuel. The Coast Guard has indicated the need for improved communication and the need to develop procedures for vessels to check in and out from fishing operations.

The Meteorological Department maintains a Cyclone Warning Division in New Delhi, with cyclone warning centres in Kolkata, Chennai, Mumbai, Bhubaneswar, Visakhapatnam, and Ahmedabad. In addition to the dissemination of warnings through traditional communication mechanisms, a new warning system using INSAT has been established.

India has all along maintained a commitment to supporting the welfare of fishers and coastal community residents through collaborative programmes of the state and union governments. These programmes address issues such as insurance, accident disability, provision of housing and water, credit for procurement of fishing equipment, and a savings scheme to provide income during closed fishing seasons.

To enhance the safety of artisanal and small-scale fishermen in India, there is the need to harmonize the resource management framework among the coastal states, develop and strengthen the enforcement of the minimum safety levels required on all vessels, and develop and enforce safety-related competency standards.

Indonesia

As a nation with some 1,700 islands, 3.1 million sq km of territorial seas, and an Exclusive Economic Zone (EEZ) of 2.7 million sq km, fisheries in Indonesia play a central role in the socio-economic fabric of the nation. Laws and regulations governing the use of marine resources have their roots in the 1945 Constitution of the Republic. There are today about 2.7 million people engaged in the primary production sector of the fisheries industry, 90% of which is small scale. The fishing fleet comprises some 0.4 million fishing boats, of which 59% are nonpowered, 24% are outboard motor-powered, and 16% are inboard motor-powered.

Traditional boats in Indonesia are mostly dugout and plank-built boats. The fishermen use a variety of fishing gear from these boats. The 4- to 5-metre long, double-outrigger canoe is used mainly in the traditional fishery. The main hull is a dugout, and sail and paddles are used for propulsion. The basic wooden hulls of the larger boats are strongly built with heavy scantlings. The main engines in many of the vessels are reconditioned automobile engines. Virtually no vessel carries any safety or fire-fighting equipment.

Regulations regarding carrying safety equipment are in place for all categories of fishing vessels based on a 1935 ordinance, revised in 1972. However, the level of enforcement is low. Monitoring and surveillance priorities focus more on fishing activities in the EEZ and poaching by unauthorised foreign vessels. Monitoring and surveillance of coastal fishing activity is absent. The Coast Guard coordinates search-and-rescue activities in collaboration with other agencies.

It is interesting to note the reasons for some of the fatalities recorded by the harbour master at the Pekalongan fishing port. They included loss overboard during rough weather from fishers sleeping at night on gill nets or seine nets stacked on the deck, fishers caught in the gear when settling seine nets, and pirates. On board one vessel, fatalities were associated with fuel blockage; the fuel tank was opened at sea, and two fishermen were sent inside to clean the tank without purging it.

Over-exploitation of near-shore resources is leading to an overall decline in the fishing economy with the resultant inability to invest in safety-related equipment or safety-oriented activity. There appears to be little skipper loyalty among Indonesian crew members. The level of enforcement of all regulations is also low.

Malaysia

The Malaysian Government has adopted a proactive stance toward sustainable utilisation of its marine resources. A strategy of maintaining the present level of production includes a programme to phase out surplus fishermen, including a scheme for buying back boats. The central fisheries management instrument is the establishment of fishing zones designed to prevent conflict between traditional and commercial fishermen and to ensure equitable distribution of resources. Prior to independence in 1958, fisheries in Malaysia were managed through a set of ordinances. The Fisheries Act was promulgated in 1963 and subsequently rewritten in 1985, with amendments introduced in 1996. All aspects of licensing, registration, and operation of small-scale fishing vessels in Malaysia fall under the responsibility of the Department of Fisheries.

Like other countries in the region, the small-scale and artisanal fisheries sector in the country was transformed by the introduction of commercial fishing gear. However, 80% of the 32,000 registered fishing vessels still operate with traditional gear, including gill nets, drift nets, hook and line, push nets, and traps.

The Malaysian fisheries monitoring, control, and surveillance system is the most comprehensively developed of all the Bay of Bengal countries. Monitoring activity covers catch and effort data, fish size, population structure, and socio-economic data regarding the fishing community. Control is maintained through a comprehensive licensing policy supported by effective legislation. Surveillance is implemented through aerial surveillance. The Maritime Rescue Co-ordination Centre coordinates search-and-rescue activity. The Meteorological Service maintains a Marine and Oceanography Division that prepares marine forecasts for the three coastal regions of the country. Forecasts are prepared twice daily, providing wind speed and direction as well as projected wave height.

Overall, there is clear evidence of a strong commitment to effective management of both coastal resources and coastal communities, including the safety of those who go out to sea. The current regulatory regime and supporting infrastructure leave few gaps as far as sea safety is concerned. However, there is a need to integrate safety issues into the regulatory framework associated with revisions to the Fisheries Act, as well as the development of appropriate life-saving appliances for carrying on traditional artisanal craft. Sea communication devices have to be expanded and the safety-training requirement integrated into fishing license renewal. There is also the need to strengthen communication, coordination, and response capability to address acts of piracy.

Maldives

The artisanal fishery in the Maldives has for centuries centred around the pole and line fishery for skipjack. The fishing boats, *masdhonis*, are typically 14.5 m long. They are traditionally built from wood and undertake day fishing activity. The three main types of fishing boats in the Maldives are the *masdhoni*, *vadhudhoni*, and *bokkura*. There are some 3,200 traditional *dhonis* and 2,000 mechanised *dhonis*. All of these are open wooden craft. *Masdhonis* are about 10 to 15 metres long overall, *vadhudhonis* are about 5 to 8 metres long, and *bokkuras* are about 2 to 3 metres long. Motorisation of *masdhonis* started in 1974, and now almost all *masdhonis* active in fishing have marine diesel engines of 22-33 hp installed on them. Motorisation of *vadhudhonis* started much later, but is now becoming quite widespread.

A *dhoni* is wide and flat forward with a short stem; it narrows a little to the aft and has a square transom stern. The fore and aft parts are decked in; about two-thirds of the length has tumble home-sides, and the well is narrow and restricted. A high transverse coaming protects the well at the fore deck. The *dhoni* is sufficiently robust and stable in local sea conditions. It maintains watertight integrity. The seaworthiness and safety of traditional Maldivian fishing crafts are well known, and boat-building experts have hailed their craftsmanship.

The fleet of 1,500 craft and 22,000 fishermen are supported by collector vessels and three island-based freezing facilities. Investments in the sector during the past two decades have resulted in the establishment of a strong technical boat-building capability and enhancements to construction and outfitting standards for the traditional vessel. Recent changes to the fisheries policy have resulted in the evolution of another class of larger vessels that engage in multi-day fishing up to 100 miles offshore. These vessels are being built both in wood and GRP without technical input relating to the design or construction standards. The fleet sector is causing considerable concern to both the Ministry of Fisheries and the Ministry of Transport in terms of sea safety.

In 1999, the Ministry of Transport and Civil Aviation, working collaboratively with the Ministry of Fisheries, Agriculture, and Marine Resources, the Ministry of Defence, and the Atolls Administration, promulgated new regulations for fishing vessels. These regulations comprise the basis for the issuance of a safety certificate, which covers communication, navigation and safety equipment, lights, and signals.

In the Maldives, the control and surveillance function focuses mainly on the outer EEZ to deter poaching by foreign vessels. The National Security Service and the Coast Guard conduct approximately 400 enforcement visits to the various islands annually. Search-and-rescue activities are regulated by the Coast Guard, which responds to nearly 200 calls each year, especially during the southwest monsoon period.

Maldives is less exposed to the severe cyclonic weather patterns experienced further north in the Bay of Bengal. Weather forecasts provide both wind speed and wave height. A 24-hour service is maintained for fishermen. Storm advisories are issued 18 hours ahead. There are no formal fishing community organisations in the Maldives through which sea safety issues can be addressed. However, the historical and administrative focal point for communities throughout the country has been the Atolls Department, which coordinates the delivery of services to the respective island residents.

While the Maldives EEZ covers approximately 1 million sq km, the artisanal fishing fleet is relatively small and homogenous in nature. With the exception of the emerging new fleet sector, the vessels fish a common fishing profile, which has few inherent risks. The fishing environment is not particularly hostile, and regular marine forecasts are available throughout the country. Maldivian fishing vessels are of a design that has evolved over centuries to most effectively address the sea conditions within and between atolls. Construction standards of the wooden vessels are high. Standards for locally built GRP vessels, however, require upgrading.

The new fleet sector is growing rapidly with no regulations in place for design approval, construction, and outfit standards. The unregulated use of propane to fuel fishing lights is a potential fire risk. The use and onboard fuelling of small gasoline engines to drive spray pumps is also a fire risk. Severe fatalities have occurred through the improper use of scuba gear when diving for lobster or supporting bait-fishing operations.

Sri Lanka

The fisheries management regime in Sri Lanka was initially an open-access regime managed under the Fisheries Ordinance of 1940. Subsequently, regulations to specific regions or fisheries, such as the beach seine or purse seine fisheries, were introduced to reduce gear conflicts. Following clear evidence that coastal fishing was occurring beyond sustainable limits, a new Fisheries and Aquatic Resources Act was promulgated in 1996. The active fishing population in Sri Lanka is estimated to comprise approximately 0.15 million fishers landing some 0.16 million tonnes annually.

The small-scale fleet in Sri Lanka is typically divided into traditional craft and introduced craft. The former category comprises approximately 14,000 dugout canoes (with or without outriggers), beach seine craft, and *kattumarans*, the majority of which are unauthorised. A number of fisheries development programmes during the past 15 years has resulted in the construction of vessels for mullet day fishing in offshore waters.

Sri Lanka's traditional craft consist of dugouts with and without outriggers, log rafts, and planked craft. It is estimated that the country has about 16,500 traditional fishing boats, of which about 1,400 are motorised. The indigenous fishing crafts represent a large number of different building methods and designs. They are mainly utilised for coastal and lagoon fishing.

The *oru* outrigger canoe has existed for centuries with almost no change. An advancement of the *oru* has been the attachment of the outrigger as a balancing device. In the early days, the *orus* were propelled only by oars. The introduction of the sail was also a big achievement. The *orus* are gradually giving way to fibre-reinforced plastic boats, which fish in deeper waters and can make trips of 2 to 3 weeks.

Under the 1996 act, the Director of Fisheries and Aquatic Resources is responsible for the licensing of fishing operations and the registration of all fishing vessels. The Ministry of Transport currently plays no role in the registration or certification of fishing vessels or operators. The 1996 act empowers the Director of Fisheries to designate fisheries inspectors and other nominated personnel as authorised officers to enforce regulations. A Coast Guard act is also being drafted. Search-and-rescue activities for the small-scale fisheries in the country have traditionally been fishing community issues, with local vessels assisting as and when required.

Two specific environmental conditions impact the safety of small-scale fishing operations in Sri Lanka. For one, many of the traditional craft are beach-landed, and for another, the country is subject to severe storms associated with the onset of the southwest monsoon. The Department of Meteorology maintains 10 observation stations around the coast and nine stations inland. Marine forecasts are prepared twice daily and forwarded to the Ministry of Fisheries and Aquatic Resources, which broadcasts them at specified times.

Although the fisheries policy environment in Sri Lanka is proactive and supported by professionals, there is an absence of succession planning. Gaps are present in terms of standards for the construction and outfitting of safety equipment for all fishing vessels. Monitoring is perceived to be weak. No standards exist for fishing vessel operator training or competency certification. There is an urgent need to develop design and safety standards for multi-day vessels and to improve communication equipment

Thailand

The fisheries policy in Thailand has evolved over the past three decades to reflect the changing circumstances of resource abundance. Under the first three National Economic and Social Development plans, the fisheries policy focused on increasing production, particularly through the exploitation of resources in the Gulf of Thailand. Under the eighth plan, the focus has been to attain fisheries sustainability with the rehabilitation of fisheries resources and environment.

The endeavour has also been to implement regulations governing the conduct of Thai fleets in compliance with fishing or joint venture agreements with other coastal states; to accelerate coastal aquaculture development with diversification of cultivable species; prevention and control of environmental degradation due to aquaculture development; and to improve the quality of Thai fish and fishery products to enable them to compete in foreign markets. The need for a new fisheries act became evident in the early 1990s, following rapid decline in coastal fisheries resource abundance. A comprehensive draft law was completed in 1996, but has not yet been adopted.

Based on a 1995 census, the profile of the artisanal and small-scale fishing fleet included 2,826 unpowered boats principally using lift nets and crab traps, 35,430 outboard-powered boats using various types of gear, 6,925 inboard-powered boats of less than 10 gross tonnage (GT), 6,550 inboard-powered boats of 10-49 GT using otter trawls or pair trawls, and 1,807 inboard-powered boats of 50 GT. Traditional fishing boats in Thailand do not vary much in shape and structure.

Fishing licenses are issued annually by the Director of Fisheries. Monitoring, control, and surveillance functions are carried out by the Fisheries Resources Conservation Division of the Director of Fisheries, which maintains seven stations in the Gulf of Thailand and four along the coast of the Andaman Sea. The Thai Maritime Enforcement and Co-ordination Centre coordinates search-and-rescue operations.

Fishermen on the Andaman Sea coast experience the effects of cyclonic activity in the Bay of Bengal, while those in the Gulf of Thailand are subject to typhoons. Marine forecasts are issued twice daily using radio and television. Storm warnings are also displayed at coastal fisheries stations. Most fishers using mechanised boats carry radios. The fishermen's cooperatives offer the best potential for promoting community participation in sea safety.

The principal factors affecting sea safety were seen to include conflicts associated with the intrusion of trawlers into zones reserved for passive gear; lack of appropriate safety equipment on all vessels; trawler designs that carry a lot of top hamper that are at risk during typhoons or cyclones; and the declining economic viability of small-scale fishing enterprises associated with a declining resource base, leading to higher levels of risk taking.

The Chennai Declaration on "Sea Safety for Artisanal and Small-Scale Fishermen" and a Summary of Factors Affecting Sea Safety for Artisanal and Small-Scale Fishermen in the Bay of Bengal region are provided in Appendixes A and B, respectively.

Conclusion

The issues highlighted in the foregoing paragraphs are not insurmountable, but may need many attempts, and perhaps a long-term programme, to inculcate the habit of sea safety among fishermen and reduce the loss of human lives and misery. Although the basic problems of safety at sea are common to all the countries of the Bay of Bengal, local conditions and complexities vary widely. Creating a safer working environment for artisanal and small-scale fishers is a huge challenge. It merits urgent and critical attention from all concerned, especially the donor countries that can provide the much-needed assistance for implementing a long-term programme on sea safety in the region.

Safety measures have to be tailored to meet the specific requirements of each country in the Bay of Bengal region, in Cupertino with policy makers, legislators, vessel owners, fishermen, and other stakeholders. The economics of safety measures should be an important criterion, and awareness of the need for a legal framework covering safety aspects is absolutely necessary.

To ensure integration of sea safety measures in the day-to-day life of the fishermen, it is essential that they be built around the entire community and not the fishermen *per se*. The school curriculum in the coastal areas should include lessons in sea safety. This would go a long way in preparing a young fisher boy to adopt safety habits when he graduates into a mature fisherman. Extension programmes are necessary for the fisherwomen so that they can persuade the men-folk to use safety measures for their protection and for the protection of the family.

Safety at sea should be an integral part of fisheries management. Implementation of sea safety programmes should include mandatory regulation, a sound implementation mechanism, training and education, prevention and survival strategies, etc. Finally, to make it sustainable, both groups, the government, and the fishermen, should share the effort and assistance.

I place on record my thanks to the Fisheries Division of the Food and Agriculture Organisation, Rome, and the Alaska Marine Safety Education Association for providing me this opportunity to participate in the conference and present this case study on the Bay of Bengal region.

Appendix A: Chennai Declaration on Sea Safety for Artisanal and Small-Scale Fishermen

Conscious that fishing is the world's most dangerous occupation with more than 24,000 deaths per year attributable to weaknesses in the institutional and regulatory environment, a declining resource base, and poor socio-economic conditions in the sector;

Realising that sea safety regimes are weakest amongst the artisanal and small-scale fisheries sectors, particularly in developing countries;

Realising that more than 80% of the world's artisanal and small-scale fishers are concentrated in Asia, where many of the coastal target stocks are over or fully exploited;

Recognising that the consequences of loss of life fall most heavily on the surviving families, for whom alternative sources of livelihood may not exist;

Concerned about the inadequacy of social and political will to address the issue of fatalities among artisanal and small-scale fishermen;

Accepting that the issue of safety for the artisanal and small-scale fisheries sectors is not fully recognised, or acknowledged, by fisheries policy objectives and further, that the focus is more on economic and resource management issues than the safety of artisanal and small-scale fishermen;

Concerned that current fisheries management regimes for coastal fisheries in the region may lead to increased levels of operational risk for artisanal and small-scale fishermen;

Concerned that safety measures, together with supporting regulations and standards relevant to the needs of artisanal and small-scale fisheries sectors, remain inadequately addressed by fisheries and maritime administrations in the region;

Recognising that neither the Torremolinos International Convention for the Safety of Fishing Vessels, 1977, as amended by the 1993 Protocol, and the

1995 Convention for the Standards of Training, Certification and Watch Keeping for Fishing Vessel Personnel are in force, nor are they applicable to fishing vessels under 24 metres in length;

Recognising the limitations in institutional capacity of fisheries and maritime administrations in the region to undertake all responsibilities associated with their mandate;

Realising that fishing operations are carried out in a hostile and hazardous environment from vessels often having weaknesses in their design, construction, and equipment, thus being prone to failure;

Accepting that fishermen in both traditional and diversified fisheries are exposed to inherently high levels of risk and resulting accidents, for which there are few survival or rescue strategies;

Emphasising the urgent need to address the multi-dimensional issue of sea safety for artisanal and small-scale fishermen on a regional basis and in a holistic manner; and

Recognising that the problem is not insurmountable;

We, the representatives of Fisheries and Maritime Administrations, Coast Guard/Navy and Fishermen's Associations, nominated by the Governments of Bangladesh, India, Indonesia, Malaysia, the Maldives, Sri Lanka, and Thailand, having participated in the BOBP/FAO Regional Workshop on Sea Safety for Artisanal and Small-scale Fishermen held in Chennai, India, from 8 to 12 October 2001, now therefore:

Resolve to address, as a matter of urgency, the issue of safety at sea for artisanal and small-scale fishermen;

Recommend that sea safety issues be comprehensively integrated into member countries' fisheries policy and management frameworks. This would include associated commitments under the Code of Conduct for Responsible Fisheries and other regional, inter-regional or global instruments and initiatives;

Recommend measures which would result in a harmonised and holistic fisheries management framework for the Bay of Bengal;

Emphasise the need to rationalise institutional mandates, legislation, regulation and enforcement at the national level, in order to enhance sea safety in artisanal and small-scale fisheries;

Ensure the incorporation of FAO/IMO/ILO Voluntary Guidelines for the Design, Construction and Equipment of Small Fishing Vessels and the FAO/IMO/ILO Document for Guidance on the Training and Certification of Fishing Vessel Personnel into regulatory frameworks, as appropriate;

Recommend that fisheries and maritime administrations enhance their knowledge of the operations and constraints of the artisanal and small-scale fisheries sectors in order to formulate effective guidelines, standards, and regulations for the safety of fishing vessels, including the certification and training of crews;

Recommend the development and implementation of education, training and awareness programmes, which satisfy regulatory requirements, while also building a culture of sea safety within artisanal and small-scale fishing communities;

Recommend that mandatory requirements for improving sea safety be supplemented by other strategies, which involve the participation of the fisher communities, families, the media, and other stakeholders in order to promote the adoption of a wide range of safety measures;

Recommend that member countries undertake measures directed towards ensuring enhanced economic viability of artisanal and small-scale fishing enterprises as an essential element of the sea safety issue;

Recommend that administrations consider the provision of financial and other incentives to encourage and ensure the widespread use of safety equipment, together with training in the use of such equipment;

Recommend that a programme of applied research and development be initiated, focusing on the development of cost effective safety related equipment relevant to the needs of the artisanal and small-scale fisheries sectors;

Strongly recommend the formulation and implementation of a regional sea safety programme, employing a consultative and participatory approach, building upon institutionally derived data, together with the operational experience of artisanal and small-scale fisher communities;

Recommend that the issue of sea safety be addressed on an urgent basis, possibly achieved through a regional mechanism such as the Inter Governmental Organisation proposed by the BOBP member countries during the 24th meeting of the BOBP Advisory Committee at Phuket, Thailand (The Phuket Resolution - October 1999);

Agree to seek the support of the donor community for the development of a sea safety programme, and also request FAO to seek such assistance on our behalf.

(Adopted on Friday, 12 October 2001 in Chennai, India)

Appendix B: Factors and issues of concern in Bay of Bengal fisheries

Country	Factors	Issues of Concern
Bangladesh	Open-access mangement regime leading to overexploitation, reducing overall profitably of operations; affects investing in safety-related equipment or safety orientation activities. Need to coordinate and harmonise the regulatory environment. Need to upgrade mechanical installations to replace use of tube well engines. Upgrading communication equipment on artisanal vessels. Risks associated with the economic structure of the industry. General need to upgrade fishermen safety training and awareness.	Weak enforcement of all safety and operator competency standards. Fishing gear conflicts. Cyclone-related risks. Piracy.

Country	Factors	Issues of Concern
India	Need to harmonise resource management framework between respective states with a view to reducing conflicts amongst adjacent resource users. Need to develop and strengthen the enforcement of the requirement to carry a minimum level of safety-related equipment on all vessels. Develop low-cost safety equipment for use on small-scale vessels. Need to promote expanded use of communcation equipment at sea, together with training in its proper use. Increased dialogue between Sri Lanka and India with a view to reducing unacceptable levels of risk and detainment of fishers. Continuing assessment of resource management instruments to determine their impact on socioeconomic structure of coastal communities and associated levels of safety. Potential for development of informal, community-based search-and-rescue activities in an auxiliary Coast Guard model. Attention to communication and community participation in disaster prevention.	

Country	Factors	Issues of Concern
Indonesia	Review of sea-safety-related incidents indicates that the safety issue must be addressed primarily through the window of practice and awareness and secondarily through the window of legislation and regulation. The Fisheries Department is in a state of transition and there is a lack of clarity in functional responsibilities with adverse impacts on monitoring, control, and surveillance activities. Overexploitation of near-shore re-sources reduces overall profitabililty of operations, affects investing in safety-related equipment or safety orientation activities. Crew move from boat to boat, making it difficult to develop integrated team approach at sea with an associated adverse effect on safety issues. The national network of fishermen's associations offers promise as a ve-hicle to upgrade safety standards and practices throughout the sector. Practically oriented education and training capability is in place.	Levels of enforcement of all regulation low. Vessel outfitting standards on all vessels dangerously low. Low level of adoption of technology in navigation or communication equipment.

Country	Factors	Issues of Concern
Malaysia	Clear evidence of a strong commitment to effective management of both coastal resources and coastal communities, including the safety of those who go to sea. The current regulatory regime and supporting infrastructure leave few significant gaps in the sea safety question. Integrating safety issues into the regulatory framework is underway. Developing appropriate life-saving appliances for use on traditional artisanal craft. Strengthening the regulatory requirement for life-saving appliances for use on traditional artisanal craft. Expanding the use of at-sea communication devices. Promoting a Coast Guard auxiliary model for search-and-rescue. Integrating safety training requirements into fishing license renewal.	Strengthening the communication, coordination, and response capability to address acts of piracy.

Country	Factors	Issues of Concern
Maldives	Few inherent risks in the artisanal sector. Fishing environment not particularly hostile and good forecasting systems prevail. Traditional boat building evolved to suit the sea conditions. Appropriate standards for safety equipment introduced recently. High percentage of boats carry either VHF or CB radios and GPS. Administrative structures for regulation and enforcement in place. All agencies familiar with safety issues but are under-resourced. Coordinated efforts to expand fisheries training and increase safety awareness. Few fatalities in the fishing sector.	New fleet growing rapidly with no regulations for design approval, construction, or outfit standards, nor operator certification standards. Unregulated use of propane to fuel fishing lights. Use and on-board fueling of small gasoline engines to drive spray pumps. Improper use of scuba gear.

Country	Factors	Issues of Concern
Sri Lanka	Proactive policy environment, supported by widely experienced professional and technical capability. Safety is recognised as a serious policy issue. Community-based resource management models are being pilot-tested and evaluated. Monitoring in regulation of fishing effort and activity appears to be weak. No standards for fishing vessel operator training or competency certification. Need to strengthen monitoring, control, and surveillance capability. Search-and-rescue capability most effective at the community leve. Environmental forecasting capability sound and effectively distributed. Increased conflicts between traditional and introduced vessels.	Need to develop and design safety standards for multi-day vessels. Need to improve the engineering and navigation skills of operators of multi-day boats. Need to improve level of communication equipment, operator training, operator radio discipline. Assign emergency radio channel for fishermen. Need to channel safety training and awareness through community organisations. Need to address issues associated with apprehensions of fishermen in neighbouring countries.

Country	Factors	Issues of Concern
Thailand	Conflicts associated wth intrusion of trawlers into zones reserved for passive gear fishers. Lack of appropriate safety equipment on all vessels. Trawler designs that carry a lot of top hamper, which are demonstrably at risk during extreme typhoon or cyclonic weather conditions. Piracy. Declining economic viability of small-scale fishing enterprises associated with declining resource base leading to higher levels of risk-taking.	

SESSION SEVEN:
VESSEL SAFETY

African fishermen *(Photo courtesy of Ousmane Ndiaye)*

Session Seven

OPERATION SAFE CRAB:
A RISK-BASED REGIONAL INTERVENTION

Kenneth M. Lawrenson
Commercial Fishing Vessel Safety Coordinator
U. S. Coast Guard Marine Safety Office
Portland, Oregon, USA

Curtis J. Farrell
Commercial Fishing Vessel Safety Examiner
U. S. Coast Guard Marine Safety Office
Portland, Oregon, USA

Daniel E. Hardin
Commercial Fishing Vessel Safety Coordinator
U. S. Coast Guard Thirteenth District
Seattle, Washington, USA

The views expressed herein are those of the authors and are not necessarily the views of the US Dept. of Homeland Security or the US Coast Guard.

Any investigative report regardless of its completeness and soundness, aside from good reading, has little or no value, unless immediate action is taken to implement reasonable and sound recommendations made by the board to prevent a repetition. (Captain Dominic Calicchio, US Coast Guard, Retired, on the apparent lack of action after the sinking of the SS *Marine Electric* in 1983, with a loss of 31 crew. Quoted in *Until the Sea Shall Free Them*, by Robert Frump, 2002.)

Identifying a high-risk fishery can be subjectively easy, even though reliable statistical data are often difficult to obtain. Blending casualty, population, and environmental data, along with a healthy dose of best guess, the US Coast Guard's Thirteenth District (see note a) recognized the vital need to develop and implement an at-the-dock safety intervention for Oregon and Washington commercial Dungeness crab-fishing vessels. A tragic series of mishaps during the 1999-2000 season provided the most compelling reason to act and prompted Operation Safe Crab for the last three crab seasons.

Operation Safe Crab is a bold idea for the Coast Guard: An attempt to replace previous random voluntary dockside safety examinations with a targeted, large-scale, on-the-dock Coast Guard presence tied to credible consequences for those vessels unable or unwilling to comply with federal safety regulations. The authors will present underlying data and analyses that support a risk-based approach to improving safety for this fishery. In addition, we will tell the story of the goals, planning, resistance within the Coast Guard to this effort, efforts to ensure the program and related consequences are legal, deployment of resources, cost count, and examination of the results of our efforts.

Hazards of the Dungeness crab fishery

The Oregon and Washington commercial Dungeness crab fishery is extremely dangerous. The season typically opens December 1, although market forces and the state of the crab's hardness and meat content (resulting from natural molting cycles) have delayed the actual start of fishing more often than not in the past 10 years, sometimes by as much as 3 weeks. The hazards arise from several sources.

- Winter presents the worst weather conditions of the year on the Pacific Ocean off Oregon and Washington.
- Oregon and Washington coastal ports are located on river entrances with hazardous bars. Although the Columbia River Bar has perhaps the most notorious reputation (i.e., the "Graveyard of the Pacific"), the bars at Gray's Harbor, Tillamook Bay, Yaquina Bay, Umpqua River, Coos Bay, and the Chetco River can be equally, if not more, treacherous to navigation. Local Coast Guard personnel know from anecdotal evidence that mishaps during hazardous bar crossings account for about two-thirds of the commercial fatalities.
- There is intense pressure to fish: Crab income generally represents a significant part of annual income for typical vessel owners. Adding to this pressure is the fact that although the season remains open until August, 75% to 80% of all Dungeness crab is landed during the first 2 months of season opening. In addition, the holidays are seen as both a reason to deliver product for the lucrative holiday market and as a source of cash to meet holiday expenses. The end result is a fishery that has all of the "race to fish" aspects of any derby fishery.
- Dungeness crab is a pot fishery in which vast numbers of pots are placed in the relative shallows, usually at depths of less than 50 fath-

oms. This places the crabber in close along the coast where surf conditions can be at their worst. It also requires the vessels to travel at
low speed, with working gear over their sides, on along-shore course
headings that place them beam to the prevailing swell and subject to
the greatest potential for rolling. A worse set of operating conditions
is difficult to imagine.

* Crab vessels themselves tend to have poor stability characteristics
when loaded with pots. Although some vessels are able to load part
of their gear into the hold where the low center of gravity maintains
adequate stability, the standard practice is to load pots on deck until
reaching a pot load that the vessel owner or operator "knows" from
personal experience will be "safe." These deck loads, so necessary
to ensure large crab landings, inevitably raise the vertical center of
gravity and reduce intact stability (see note b). Exacerbating this
reduction of stability is the dangerous free-surface effect from liquid
loading in the crab holds and frequent water on deck (see note c).

All these factors result in frequent tragedy. Those who push hard and are
lucky make good money. Those who push hard and aren't lucky have a bad
day and cause the Coast Guard to get involved.

Casualty history

The last seven Dungeness crab seasons off Oregon and Washington claimed
the lives of 16 men (Table 1) (see note d).

Calculations involving degree of risk require a denominator, such as number
of vessels, number of fishermen, or number of hours operated. Several data
sources are typically used, with varying degrees of manipulation and assumptions. Such normalizing of data has been, and continues to be, a difficult
issue (US Coast Guard 1999).

Two figures are needed to assess this fishery's risk and the extent to which
Operation Safe Crab impacted the fleet: The total number of vessels participating in the season and the number of fishermen working in this fishery.
Data compiled from the states of Washington and Oregon by the Pacific
States Marine Fisheries Commission's PACFIN database were used to
analyze the number of crabbing vessels in the fishery. Actual crab landing
information was aggregated by individual vessel for Dungeness crab landed
in either Oregon or Washington. Figures for total weight of crab delivered,

Table 1: Casualty history, 1996-2003

Crab season	Vessels with fatalities	Total season fatalities
1996-97	Beach King (1 death)	1
1997-98	*Jolly Roger* (2 deaths)	
	Seeker (1 death)	3
1998-99	None	0
1999-2000	*Blue Heather* (2 deaths)	
	Silver Spray (1 death)	
	Paula C (3 deaths, see note below)	6
2000-01	*Miss Brittany* (2 deaths)	2
2001-02	*Nesika* (4 deaths)	4
2002-03	None	0
Total, 1996-2003		16

Note: The *Silver Spray* and *Paula C* were actually lost off northern California, but the circumstances and geographic proximity are so similar to those accidents off Oregon that they are included in this table.

as well as number of trips and total crab revenue, were then totaled by crab season, rather than by year. The season was defined as the fourth quarter of a year plus the first, second, and third quarters of the following year.

Any crab vessel making at least one commercial landing in that season was included in the list (see note e). However, a Pareto analysis of the "weight landed" data showed that an overwhelming percentage of the vessels on the list landed so few crab that they could not be truly considered "commercial" in the sense we were after—vessels that carried wage-earning crews and that fished the season with the intention of commercial success. For example, in the first quarter of 1997, there were 810 vessels that made at least one commercial crab landing. Half of that number, when rank-ordered by crab weight, were responsible for landing 90% of all crab in that period. Clearly, many permit holders are not commercially viable. To approach a more realistic number of commercial vessels, two cut-off points were examined. The first was the number of crabbers that, when rank-ordered by crab weight landed, delivered 90% of the total amount of crab landed. The second was the number of crab vessels that landed at least 20,000 pounds of crab during the season; this figure represents estimated minimum revenue that would allow payment of meaningful crew earnings (Table 2; Figure 1).

Data for fishing employment were obtained from Woodley (2000, see note f), and an estimate of full-time employees (FTEs) for the Oregon and Washington Dungeness crab fishery was made from the method described by Woodley. The method involves determining an average crew size for a vessel, multiplying the average crew size by the days operated (Table 3), dividing by 365 (days per year), and multiplying by the number of vessels in the fishery. The result is an estimate for annual employment equivalent (AEE). Then a

Table 2: Landings

Crab season	No. of vessels with at least 1 landing	No. of vessels at 90% cutoff	No. of vessels landing 20,000 pounds or more
1996-97	1457	586	306
1997-98	1537	630	342
1998-99	1521	554	337
1999-2000	1516	501	426
2000-01	1538	659	349
2001-02	1546	597	418
2002-03	1330	440	440

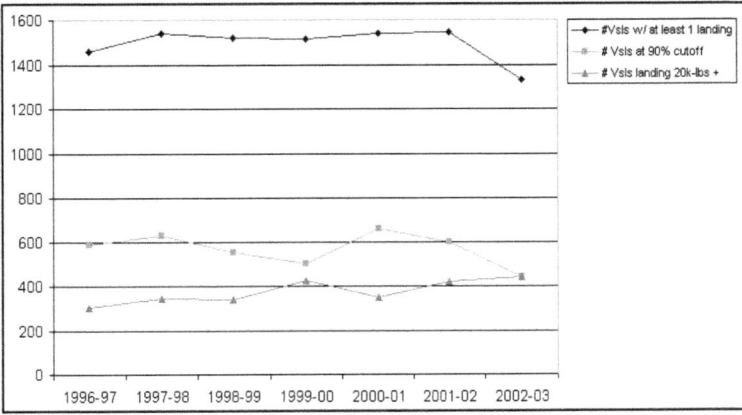

Figure 1: Landings

Table 3: Average number of operational days per vessel

State	1996	1997	1998	1999
Oregon	77	65	62	78
Washington	109	113	136	117

conservative factor of three is used to convert this equivalent to a risk exposure, since the AEE is based on an 8-hour workday, whereas commercial fishermen are typically aboard the vessel (and so subject to vessel casualties) for the entire 24-hour period of that day. Since the number of crab vessels was not divided evenly between Oregon and Washington, an assumption was made that the two figures are roughly equal. Therefore, a straight average for operational days for the entire region was used.

AEE for risk exposure = 3 × (average no. of operational days) × (crew per vessel × (no. of vessels) ÷ 365

Crew size per vessel was assumed to be three for a typical crab vessel, and the number of crab vessels landing at least 20,000 pounds of product is assumed to be the average for crab seasons 1996-1997, 1997-1998, and 1999-1998, or 328. This gives an AEE of 763 for 1996, 730 for 1997, 812 for 1998, and 804 for 1999, resulting in an average AEE of 777 for risk exposure over the 4 years. The total number of fatalities for this period was 10, or an average of 2.5 deaths per year.

The fatality rate per 100,000 workers is typically used to compare risk between different fisheries (or other occupations) and is calculated as 100,000 times number of fatalities divided by AEE. For the Oregon and Washington Dungeness crab fishery between 1996 and 1999, the fishing fatality rate per 100, 000 workers was 322.

This figure is twice as high as the overall national fatality rate for commercial fishing estimated for 1995 (US Coast Guard 1999). Anecdotally, and certainly in our guts, we knew there was a problem. The disaster of the December 1999 crab opening should have come as no surprise. Action was demanded.

Idea of a fisher-specific intervention

Medlicott (2002) describes a common complaint among experienced Coast Guard fishing vessel safety personnel: The voluntary dockside examination program has no teeth to compel compliance with safety regulations. At-sea boardings failed to make up for this deficiency, and no specific action was being taken toward the highest risk groups. Medlicott further described the efforts of the Seventeenth Coast Guard District (Alaska) to implement an at-the-dock effort to reach a large percentage of the Bering Sea king crab fleet during the 1999-2000 season.

Personnel from the Thirteenth and Seventeenth Districts involved in fishing vessel safety have enjoyed a close partnership since the inception of the regulations in the early 1990s. When word spread of the successes by the Seventeenth District, it was obvious that a similar at-the-dock enforcement action could have an impact in the Oregon and Washington Dungeness fishery. Several key factors were present.

- The majority of the vessels involved would be departing from a relatively small number of ports. This meant it would be possible to focus personnel resources on only the ports where the crabbers were actually located. In other words, we would know *where* to find them.
- The majority of the vessels involved would be departing at roughly the same time, a time that would be predictable and was driven by fishery management. We would know *when* to find them.
- The start of the crab fishery is labor intensive: crab pots and other gear are to be repaired and loaded on board, bait prepared, holds are to be inspected by the state. This would mean that vessel crews would be on the docks, working on their vessels, immediately before the opening day. We would know that the *crews* were there.

While there was support for an operation for the 2000-2001 season from the Command at the Coast Guard Marine Safety Office in Portland, Oregon (in whose area of responsibility the bulk of the effort was to take place), there was resistance from several quarters within the Thirteenth District (see note g). Several of the coastal units were worried that such an effort would damage the relationship between the Coast Guard and the fishing fleet, because the Coast Guard had never before threatened to prevent a vessel from fishing until it came into compliance. The District Command was concerned that any actions taken be completely within existing authority and jurisdiction. Lastly, concern was expressed that the industry needed to know well in advance what we were going to do, what they needed to do in order to "pass" our examination, and what possible consequences could befall those failing to comply.

Clearly, the challenge would be in the planning. Thorough planning would sculpt a "stolen" idea into Operation Safe Crab. It had to start with the goals.

Goals of Operation Safe Crab

Eliminating commercial deaths in the fishery by increasing vessel safety was the ultimate goal. To accomplish this, we had to be more specific in our objectives. These were to—

- Examine as many Dungeness crab vessels as possible before the season.
- Focus attention on these most critical issues given the risks inherent to the fishery.
- Keep examinations as short as possible to accomplish our job while minimizing impact to a fleet busy getting ready for the opening.
- Ensure consistency of enforcement.
- Keep the examinations safe for the Coast Guard personnel involved.
- Provide a credible continuum of consequences for vessels that were not in compliance.
- Involve law enforcement personnel of the Thirteenth District.
- Ensure that the industry knew in advance what we would be doing, so that it would not come as a surprise.
- Manage limited budget resources to support the efforts.

Planning the operation

The first objective was to examine as many vessels as possible. Concentrating examiners where the vessels were going to be would do this. The "where" was easy to determine. The timing of the operation, however, became a matter of considerable discussion among the various concerned captains within the Thirteenth District. It was clear to the authors that to check a vessel effectively, the crew needed to be aboard. Therefore, to maximize our chance of meeting both crew and vessel at the same time, the Coast Guard would need to wait until the final few days immediately before the start day. In addition, the vessels were already required to be in compliance, and if boarded at sea in the same condition we found them at the dock, their voyage would be terminated without their being able to continue fishing. Others suggested that the Coast Guard should examine the vessels several weeks before the opening so that a vessel found out of compliance would have the time necessary to come into compliance before the start. In the end, however, the former argument carried the day, although widespread publicity would now become a critical part of the operation. During this discussion, it was clearly reiterated that this operation would *only* involve vessels engaged in the Dungeness crab fishery, rather than any vessel we would come across.

To focus effort on those matters most directly related to vessel safety and to keep the examinations short, three areas of emphasis were to be made: the vessel's stability, the vessel's life raft, and the vessel's emergency position-indicating radio beacon (EPIRB). The vessel's stability was seen as the primary factor preventing vessel casualties and could be appropriately gauged by examining the loading and number of crab pots, the vessel's freeboard, its compliance with stability instructions (if required), and its crab holds and deck freeing ports to ensure minimal free-surface effects. Checks of other lifesaving equipment (the raft and EPIRB) were meant to give the crew the greatest chance of surviving a vessel casualty. In previous crab vessel losses, the condition of these two pieces of gear had been found to make the critical difference between surviving and not surviving. A checklist was developed to gather some basic data on the vessels and to document the checks for these three areas. Our goal was to go on board a vessel and, if no deficiencies were noted on the checklist, leave within 10 to 12 minutes. This would allow an examiner to complete examinations on five to six vessels per hour.

The checklist would be one method to ensure consistency of enforcement. The other measure involved selecting examiners. It was clear that when doing such an operation for the first time, we needed to use the right people. Six experienced, seasoned examiners were chosen and paired into three teams. Each team leader had been a key planner in the operation, and each had an excellent perspective of the objectives and methods.

To ensure the safety of the Coast Guard examiners, several guidelines were adopted. First, no vessel would be boarded without a crew member being present. Vessels would be examined at dockside only. Vessels would only be examined during daylight. Lastly, each team would check in with the Coast Guard station in that port so that if problems were encountered with a vessel, Coast Guard law enforcement personnel would be available as back-up.

Providing examiners with credible consequences for noncompliance necessitated cooperation and communication with the Captain of the Port (COTP) and the District's Law Enforcement Division. At the very least, vessel deficiencies would be recorded and transmitted to the district at the end of each day, with a final record of vessels in noncompliance made available to all Coast Guard units in the district to help prioritize boardings. Vessels with more serious problems, such as problems that would result in voyage termination if detected at sea, were to be issued a COTP order directing that vessel to remain in port until the deficiency was corrected.

Publicity was critical and started about a month before the projected start of the season. Press releases were sent to coastal radio stations and newspapers, flyers were posted at marinas, and fishing associations were contacted so that the information could be passed on to their membership. This was to be no secret.

The Thirteenth District commander approved the plan. The key aspects were released in message form to all the Coast Guard units involved.

Operation Safe Crab 2000

Originally, we planned for three teams, each assigned an area: north, central, and south. The North Team would examine vessels in the Columbia River ports of Astoria, Warrenton, Hammond, Chinook, and Ilwaco; Willapa Bay to the north of Ilwaco, Washington; and Westport, Washington, on Gray's Harbor. The Central Team would cover the ports of Florence, Newport, Depoe Bay, and Tillamook Bay (all in Oregon). The South Team would be responsible for the ports of Brookings, Port Orford, Coos Bay/Charleston, and the Umpqua River (Winchester Bay and Reedsport) in Oregon.

At the last minute, the marketability of the crab required the season to be split, north and south; the south opened on December 1, but the north was delayed 2 weeks. Rather than remain with the plan and chance missing many vessels, we decided to be flexible and split the teams along the same geographic lines. The North Team would become a second Central Team, and we would send more examiners out to the north 2 weeks later. Fortunately, the Thirteenth District provided the additional funds.

The teams gathered their gear and traveled to their starting locations November 26. Each team had been provided with supplies of checklists, sample

COTP orders if needed, a folder containing pertinent law enforcement intelligence information, daily examination log sheets, stability and regulation pamphlets, and shackles and weak links for correcting life raft installations on the spot.

Examinations were done between Brookings and Depoe Bay on the following 4 days (November 27 to 30), with each team traveling as necessary to maximize time spent looking at crabbers. Similarly, two teams left on December 10, one to Westport, and the other to the Columbia River ports to conduct examinations from December 11 to 14. A total of 266 vessels was examined. Using 20,000 pounds as the cutoff for vessels fishing in the 2000-2001 season (349), we estimate Operation Safe Crab 2000 reached about three-quarters of the fleet. This wildly exceeded our expectations.

An alarming percentage of EPIRBs and life rafts were found to have problems. Almost 43% of EPIRBs were not in compliance, and 10% had problems so serious that the vessel would have been ordered back to port if boarded at sea. Of all life rafts on board vessels, 35% were not in compliance, and 22% were installed incorrectly. Life raft installation problems can prevent the life raft from being deployed when needed, which could lead to fatalities. Fortunately, installation issues are almost always easily corrected on the spot. One vessel was resistant to servicing its life raft (almost 2 years past servicing date), and a COTP order to correct the problem prior to sailing was issued.

When discrepancies were compared to the vessel's dockside exam status, an important conclusion was drawn. Even though almost 70% of the vessels examined during the operation had either a previous voluntary dockside exam or had been issued an examination decal (demonstrating full compliance with the regulations at the time of the exam), little difference was seen in the raft and EPIRB discrepancy rate when compared to vessels that had never been voluntarily examined by the Coast Guard. An early suggestion during the planning phase was to skip those vessels with a dockside exam decal. However, this idea was dropped because we really didn't know what the examination teams would find during the operation. Luckily, the data validated our gut feeling that our efforts should focus on all crab vessels and not just on vessels without decals.

Operation Safe Crab 2001

A similar process was followed the next year; however, the season was not split. Three teams were deployed for a 4-day period, and 148 crab vessels were examined. No COTP orders were required that year. The number of vessels with serious problems decreased by 20% from 2000 to 2001.

Anecdotally, we heard from the major marine equipment supplier that its life raft servicing workload had shifted from a pattern of raft servicing in the spring as fishermen got ready for the summer troll fisheries to one where the rafts were brought in in October and November so that they would be in compliance when the Coast Guard came out before the crab season. In addition, the supplier reported selling out of new EPIRBs in early November and had to backorder safety items.

Eighty-two of the vessels examined in 2001 were repeat exams from 2000, meaning a total of 332 individual crab vessels were examined during the 2-year period. We believe that we examined nearly every active commercial Dungeness crab vessel for an expenditure of 64 man-days, about $8,000 in travel-related costs. On average, each examination cost the Thirteenth District about $20 in travel expenses when comapred to the cost of at-sea boardings, for which the average cutter-day may yield three or four boardings at an operating cost of several thousand dollars per day. Thus, there was little doubt that Operation Safe Crab was saving the taxpayers a lot of money.

But was it saving lives?

Operation Safe Crab effectiveness

The authors had hoped to use Perkins' methodology (Perkins 1995) to show statistical significance by testing a 2-by-2-column table using Fisher's exact two-tailed test. Table 4 was constructed for the 2000-2001 crab season using the total number of vessels (349) from the 20,000-pound cutoff. Only one vessel examined, the *Miss Brittany*, subsequently suffered a fatality during the season. Using the STATCALC module of the Centers for Disease Control and Prevention Epi Info software package proved moot. Due to the lack of differences between examined and unexamined vessels, the p-value was 1.000. This is true also for the 2001-2002 and 2002-2003 crab seasons, since no fatalities occurred aboard crabbers we didn't examine.

Table 4: Vessel fatalities

	Vessels with fatalities	Vessels without fatalities
Vessels examined	1	265
Vessels not examined	0	83

In terms of number of fatalities, deaths decreased, from 9 to 6, in the 3-year period before and then during Operation Safe Crab. Likewise, the number of vessel casualties that resulted in a death decreased from 5 to 2 in the same period. Given a relatively flat trend in crab vessel numbers over these 6 years, these decreases are significant.

Regardless of the numbers–and quantities are admittedly small–the Coast Guard received a considerable amount of praise from the industry. Feedback from individual owners, operators, and crewmen indicated an increased level of safety awareness among the fleet. Even if we weren't changing their operational behavior, we were at least providing an additional safety net by way of enhancing the material condition of the vessels' EPIRBs and life rafts.

The remainder of this work will focus on the specifics of Operation Safe Crab 2002, with the aim of providing advice and lessons learned to other regional regulatory bodies with safety oversight duties for high-risk fisheries. May they fare as well as we have.

Operation Safe Crab 2002 in depth

Introduction

Having conducted Operation Safe Crab for 2 years, we had a solid base for conducting the next operation. The experience gained from the previous years would make Operation Safe Crab 2002 a success. This section will show in detail how we conducted the 2002 operation. We will discuss how we met and accomplished each challenge during the multi-week operation, with what went right, and what went wrong. A discussion of pitfalls and process improvements necessary for future operations of this type will complete this section.

For those who are not familiar with the make-up of Coast Guard commands along the coastal towns of Oregon and Washington, a quick description is prudent. A Coast Guard Marine Safety Office (MSO) is responsible for overall safety in the marine environment, i.e., vessels, facilities, and personnel. Some specific duties for an MSO include inspecting, examining, and certifying merchant vessels and licensing the seamen that operate them; pollution response; port security; contingency preparedness; and investigating casualties. The MSO in Portland is responsible for the coastal area from Brookings, Oregon, to Westport, Washington, and had operational control (OPCON) for Operation Safe Crab 2002. A Coast Guard group oversees search-and-rescue operations for several Coast Guard stations (both air and small boat), and a Coast Guard station operates the small boats directly responsible for conducting search-and-rescue operations. The Coast Guard groups involved in Operation Safe Crab 2002 with their stations include—

Coast Guard Group, North Bend, Oregon
- Station Chetco River, Harbor (near Brookings)
- Station Coos Bay (actually in Charleston)
- Station Umpqua River, Winchester Bay
- Station Siuslaw River, Florence
- Station Yaquina Bay, Newport
- Station Depot Bay

Coast Guard Group, Astoria, Oregon
- Station Tillamook, Garibaldi
- Station Cape Disappointment, Ilwaco
- Station Grays Harbor, Westport

Operation Safe Crab 2002 was conducted over two separate weeks due to the different season openings north and south of Port Orford. The majority of the Oregon and Washington fleet operates north of Port Orford, Washington, so most of the resources were concentrated in the northern areas. Five teams worked from the south: (1) Brookings, (2) Coos Bay, (3) Newport, (4) Astoria, and (5) Westport. These towns had large commercial fishing fleets and were close enough to the smaller ports to allow short drives to accomplish operational goals. The Brookings (south) team started the week before Thanksgiving because of the different season opening date, while the four north teams started the week after Thanksgiving.

Briefings with local units
Personal notification to each local Coast Guard station did not happen the same way at each port of the operation for various reasons with various results. Most stations received an informal in-person briefing, whereas at least one unit received no briefing at all.

Three teams met briefly with the respective local Coast Guard station representatives on Monday morning just prior to the start of operations. They were supportive and offered administrative help. They did not assist in the actual operation. This procedure worked well for these three teams.

A fourth team attempted to meet with the local Coast Guard representative at the local station and group. Both attempts failed as the representatives were unavailable or too busy at the time. This led to problems later. The local Coast Guard units had not read the operational order and without the briefing had no clear picture of Operation Safe Crab or their role in its operation.

A representative from another group actually led one of the teams. This was very advantageous, since the team leader was able to brief his commander at his leisure prior to the start of the operation. The other teams had to travel into town and then try to arrange for a briefing prior to starting Monday morning. One local station from this group was not briefed. The team leader did not anticipate needing any support from that station.

For the most part, the informal briefings worked. However, there was confusion regarding the rules of engagement from the station and the group that had not been available for a briefing. Since that station and group had not been briefed, nor had they read the operational order, they didn't understand the operational goals or the legal discussions leading to the specific rules of engagement. Therefore, future operations should be better described to the local Coast Guard representatives. Though an operation order was sent out, an in-person briefing is the best way to ensure that all participants are clear about their duties and responsibilities.

Communications

We had to keep 11 people from five teams across more than 400 miles of coastline in close communication for incident, daily, and weekly reporting requirements. This wasn't overly difficult with today's technology. The overall flow of communication was good. The following will discuss the different modes of communication used, their advantages, and their disadvantages.

- *Landline phones:* We were able to use phones at the local Coast Guard stations or groups as necessary. These phones are fixed, meaning mobility is a problem, but they are good for long calls, negotiations, or when desk space is needed while communicating.

- *Fax:* The fax was a good tool for issuing COTP orders. We used fax machines at the local stations. The field team leader would draft the COTP order, fax it to the MSO, and wait for a reply. The COTP order was then smoothed, signed by the COTP, and faxed back. The fax copy was then issued to the operator of the subject vessel.
- *Cell phones:* All team leaders carried cellular phones. Naturally, this was the quickest means of communication, and coverage was adequate. A list of all cell phone numbers was given to each team leader. The cell phone was a very good tool, allowing quick answers to questions. Persons with the right answers were always readily available so that no one was left waiting on the dock, unsure of a course of action.
- *Message system:* The Coast Guard message system was used by District 13 to report the status of vessels examined each day to the Coast Guard cutter fleet. This worked quite well, with the result that the cutter fleet had up-to-date information on the vessels they were seeing at sea. This information could then be used to prioritize boardings.
- *Internet:* Remote access to the Coast Guard Intranet user accounts proved difficult. Pulling account information from the unit to the remote unit where we were located was extremely slow and therefore not utilized. Having better remote access to computer accounts would have made some things easier, such as drafting and transmitting the COTP orders.

Personnel

The personnel used for this operation were specifically requested by name. We received excellent personnel support from many units, resulting in very high-caliber examiners for this operation. The experience level was impressive. One of the petty officers sent by a local Coast Guard cutter was its leading boarding officer. Seven people had more than 10 years of experience each as marine inspectors. MSO Portland provided two GS-12s, one warrant officer, and one petty officer. The Coast Guard cutter *Steadfast* provided one petty officer. Group Astoria and Group North Bend each provided a junior officer (ensign and lieutenant junior grade), while Coast Guard District 13 provided the bulk of the personnel with two lieutenant commanders, two GS-12s, and one GS-13 from the Seattle area.

Operational control (OPCON) was the Commercial Fishing Vessel Safety Coordinator from MSO Portland. Team leaders were Brookings - GS-12 from Portland; Coos Bay–GS-12 from Portland; Newport–GS-12 from Portland; Astoria–LTJG from Group Astoria; Westport–GS-13 from Seattle. (One of the examiners from Portland was able to lead two teams because the South Team started 2 weeks earlier due to an earlier opening day for crab season south of Port Orford.)

Walking the docks/rules of engagement

Determining the rules of engagement was a long, difficult struggle. Some Coast Guard members thought we had authority to board any vessel, any time, any place. As discussions developed, we learned from our legal office that we did not have such authority and could not board any vessel, any time, any place. We were told to ask permission to board the boats.

Our fears regarding denial of boarding were unwarranted, as we found that this system actually worked quite well. Most critically, the Thirteenth District allowed examiners the option to inform a vessel operator that failing to allow the Coast Guard to do a dockside safety check could result in that vessel being targeted for an at-sea boarding once the vessel was underway. We hoped not to use threatening language. In reality, we used that threat only four or five times during the entire operation; most operators were very compliant.

Authority and jurisdiction

We discussed three possible sources of authority with District 13 Legal and Marine Safety Divisions: 14 United States Code (USC) 89, 33 Code of Federal Regulations (CFR) Part 160, and 33 CFR §6. First, 14 USC §89 is the basic boarding officer authority for Coast Guard officers and petty officers. Boarding officer authority is a waiver of the requirement for a warrant to search. Obtaining a warrant is not practical at sea. And, although the authority states applicability anywhere (or everywhere), searching vessels at the dock is frowned upon since a warrant is easily obtained from a local magistrate. To date, District 13 has not given us permission to use boarding officer authority at the docks for Operation Safe Crab.

Next, 33 CFR §160 allows the COTP to exercise control over vessels to operate as directed, but does not allow Coast Guard personnel to gain access to those vessels. Finally, 33 CFR §6 (the so-called Super-6) came from the Magnuson Act, which was not designed for gaining access to vessels for enforcement of safety requirements.

We recognized that the Coast Guard does have authority to control the movement of vessels as described above using 33 CFR § 160. Once an especially hazardous condition on board a vessel is identified, then the COTP can issue a written order to keep that vessel from sailing until the especially hazardous condition is corrected. Draft COTP orders were provided for each team, including the list of especially hazardous conditions.

We provided the following recommendations for each of the teams.

Opening lines and statistical information
+ We are from the Coast Guard. We are here to conduct a spot check of your safety equipment.
+ I am (insert name here) with the Coast Guard. We are conducting exams of safety equipment prior to the beginning of the fishery. Is it okay to come aboard?
+ The Oregon and Washington crab fishery lost two vessels and four lives last year.
+ Lack of safety equipment and training with that safety equipment was the major cause of deaths for crab fishermen last year.
+ Thirty percent of life rafts are rigged such that they will not deploy if the vessel sinks.

Dos and don'ts
Do–
+ Be assertive.
+ Be friendly.
+ Be informative.
+ Be truthful.
+ District 13 has determined that we can, when necessary, mention to operators that those vessels that refuse a spot check will be placed on a list that will be used by law enforcement units for determining boarding priorities.

Don't–
+ Force your way onto a boat.
+ Claim authorities you don't have.
+ Provide false information.

Coast Guard reception on the docks

Once armed with our rules of engagement, we began meeting and talking with commercial fishermen. They were abundant on the docks, getting their boats ready for the crab season opener. We picked just the right time to catch them all. Asking permission to board per the rules of engagement was scary at first, but it worked. The reason it worked well was because most of the commercial fishermen want to be in compliance and want to be safe. Most of them gladly invited us to board their vessels. Some went out of their way to find us so we could check their safety equipment and ensure that their safety equipment met Coast Guard standards.

Of the 224 vessel contacts made during the 2 weeks, only five (2.2%), refused to allow a team member come aboard to check their safety equipment. Some of the few negative experiences include a fisherman who quizzed the examiners for 20 minutes regarding their knowledge and expertise (kudos for having highly experienced examiners on the docks) and then calmly allowed the examiners on board. Another fisherman growled that he was too busy and too tired. One fisherman restricted by a COTP order for an out-of-service life raft yelled and screamed at the examiners, but he was very cordial to the Coast Guard station personnel when clearing the COTP order.

We were very pleased with the attitudes of the majority of the commercial fishermen. Most of them are as concerned about safety as we are. We were able to do safety checks on hundreds of boats while protecting their rights of privacy.

Weather
Weather was not a factor. We had chilly but dry weather for the entire 2 weeks. Staying dry is important, considering all the data collection we were doing. Had it been wet we would have had difficulty keeping accurate records because we were using regular paper and clipboards. In the future we should improve our data collection tools in case we encounter a wet crab opener.

Timing
We were not able to find a fisherman on all of the vessels every day. Many boats had no one on board initially. We just kept going back to the same locations. Each time we returned we found new people we hadn't met yet. A week was about the right amount of time. At the end of the week we had met with the operator of almost every crab boat in each harbor.

Information gathering
The data collection forms were designed to allow information to be entered quickly. We worked in teams of two or three. For each vessel, one examiner

would do a quick safety equipment check, while the other asked some simple questions. We were usually on and off the vessel in less than 5 minutes. This was a good selling point when asking permission to board.

We gathered lots of information, but focused on three main areas: stability, life raft, and EPIRB. Data from previous years showed that drowning caused most deaths of commercial fishermen after their vessel capsized due to overloading. Furthermore, some of the life rafts and EPIRBS, two essential lifesaving devices, had gone down with the vessels because they had not been stowed properly. This has been confirmed as a problem through experience on the docks examining boats.

In 2000, 35% of the life rafts examined were found to be stowed incorrectly and would not have deployed or inflated if the vessel had sunk. The most common problem was the installation of the hydrostatic release. Older models of the Hammer brand hydrostatic release could be installed so that they kept the life raft from deploying. If the life raft never deploys, the painter is never pulled from the canister. If the painter is not pulled from the canister, the life raft never inflates. The new style Hammer has been redesigned and is easier to install correctly.

EPIRBs are another common problem, not only with stowage, but also with regard to their registration with the National Oceanic and Atmospheric Administration. Without a registration, an initial EPIRB signal received from the geosynchronous (stationary overhead) satellite can only tell responders from what hemisphere the EPIRB is transmitting. It can take up to 45 minutes for a polar-orbiting satellite to get within range, and then the satellite must receive several hits to vector in on a search area. With a proper registration, responders have a name, address, phone number, and description of the vessel, which helps significantly with initial response.

Vessel stability was not as easy to check as was the life raft and EPIRB. U.S. vessels less than 79 feet long or over 79 feet *and* built before 1991 are not required to have stability information. Vessels required to have stability information have a booklet from a naval architect with the exact number, size, and location for stowing crab pots. Confirming stability on a vessel with stability information was easy.

Unfortunately, the majority of crab boats in Oregon and Washington do not require stability information, so we had to devise a method for determining unsafe loading and stability. After considerable discussion with naval architects, we decided to apply a standard of 6 inches of minimum freeboard amidships, a down-flooding angle of at least 35°, and roll-period criteria. If a vessel met two of these three conditions, then we would not consider the vessel especially unsafe. It is important to note that this was never publicized as safe loading criteria; rather, it was reserved for examiners so they could be articulate about a vessel's condition when it was clearly not safe. We had a simple chart to assist with those determinations (Figure 2). If we had questions or concerns, we would contact the naval architects at headquarters and be prepared to provide them with more information about the vessel. With that, they could help determine safe freeboard, roll period, etc. Luckily, we found no vessels that were loaded in a questionable manner.

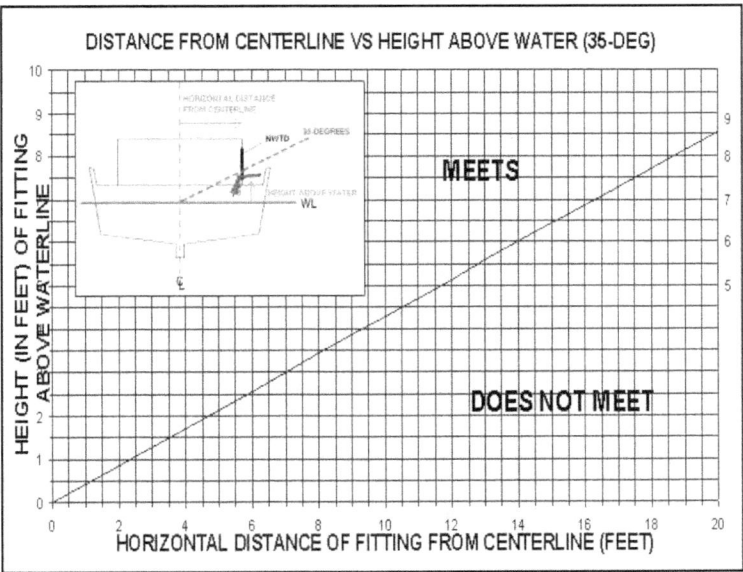

Figure 2: Safety limits

To use the chart, an examiner finds the vessel's nonwatertight fittings (i.e., pipes, hatches, or doorways penetrating the deck or hull) vertically closest to the waterline. Then the examiner measures the vertical distance from the waterline to that fitting and matches that measurement to the left side of the chart. Next, the examiner measures the distance from the fitting to the centerline of the vessel and matches that to the bottom of the chart. The numbers are then run across and up until they meet. The diagonal line on the chart corresponds to the 35° angle. Therefore, points above the diagonal line represent a fitting with a downflooding angle greater than 35° and meet the criteria, and those points below the diagonal line represent a fitting with a down-flooding angle below 35° and does not meet the criteria.

Captain of the Port orders

COTP orders are issued for vessels having especially hazardous conditions. To facilitate the decision and implement those orders, some conditions were defined ahead of time. Examiners were directed to issue a COTP order if we found any of the following conditions:

- No EPIRB on board.
- EPIRB battery more than 2 years past expiration date.
- EPIRB or life raft hydrostatic release unit more than 2 years past expiration date.
- Incorrect hydrostatic release unit rod for EPIRB (a black rod is required: older white rods have been recalled by the manufacturer).
- No life raft on board.
- Life raft capacity inadequate.
- Life raft servicing more than 1 year past expiration date.
- Life raft arrangement, unable to fix or properly stow on the spot
- Overloading.
- Lack of watertight integrity.

We were not limited by this list and could recommend a COTP order if we witnessed an especially hazardous situation not defined above.

A skeleton COTP order was provided for each team leader. If an especially hazardous condition were observed, the team leader would fill in the blanks and fax it to the MSO in Portland. The MSO would proof and smooth the document, get the COTP's signature, and fax it back. This process took a few hours. Once the signed COTP order was in hand, it was delivered to the operator of the subject vessel. The operator would then correct the especially hazardous condition and request a re-examination. Either the local Coast Guard station or one of the Operation Safe Crab team members could do the re-examination. A COTP order rescission letter was then issued to the operator.

We only issued two COTP orders. Both were for life raft servicing.

Reports

The team leaders made reports daily to the operational commander. The operational commander would combine the reports and submit them to the Office of Law Enforcement at the Thirteenth District. The report included the names of vessels contacted, number and type of discrepancies noted, if any, and/or whether the team had been refused permission to board. The daily reports were combined for a final report when the operation was complete.

Media

A media information bulletin was released several weeks prior to the operation. Local radio stations and newspapers announced the program, and at least one television station covered the information with an interview of one of the team members on the docks. Media coverage was positive.

Several commercial fishermen knew we would be on the docks because of the media coverage. They either waited for us to walk by or called and requested an examination. Our public affairs efforts were very effective.

Good press releases should always be a part of this operation.

Pitfalls

Turf battles

One Coast Guard station fought a bit of a turf battle because it felt the Operation Safe Crab team members were intruding. The team was initially treated with indifference. The station had not read the operational order and would not see the team members for a briefing. Then, as the operation continued, personnel from that station verbally challenged the authority of the team and expressed outrage at the rules of engagement that the Thirteenth District had approved. This station put its own armed personnel on the docks talking to and gathering information from the commercial fishermen and put one of their personnel with the Operation Safe Crab team. This caused confusion with the fishermen and an air of anger and distrust between the team and the station personnel. The station initially refused to provide any assistance to the Operation Safe Crab team. They later provided minimal support and then complained about it. The station complained to its group commander that the team was not dressed properly, although team members were dressed in accordance with the Coast Guard Commandant's policy for commercial fishing vessel examiners. Their group commander backed them up. The issue was never properly resolved and could lead to problems for future operations at that location.

Seven other stations were involved, all of which provided excellent support for Operation Safe Crab.

Disgruntled fishermen

The few disgruntled fishermen we met were handled successfully. According to our rules of engagement, we were to attempt to board vessels assertively. If a fisherman got angry and disagreeable, we would back off, well shy of becoming aggressive. Examiners did a good job not crossing that line.

When an attempt to board failed, we would explain to the operator that the vessel name would be placed on a list that the offshore Coast Guard cutters would be given and that could lead to being targeted for an at-sea boarding.

Issuing a COTP order was tough. Telling a fisherman he couldn't go fishing is a difficult task. We issued two orders during the operation. The vessel operators cleared both relatively quickly. In one case, the local Coast Guard station cleared the discrepancy after the Operation Safe Crab team had issued it. That worked well, as the operator, who was very angry with the issuer of the COTP order, was able to clear it with other Coast Guard personnel with whom he wasn't so angry.

There is no good way to handle someone who is yelling and threatening you, especially if you have no compliance authority or tools (weapons). However, I believe dealings with angry fishermen in future operations should be handled similar to past operations. Be calm, listen, state the facts, and leave if things get out of hand.

Cutter follow-ups

Reports included information on vessels not in compliance, vessels in full compliance, and vessels whose operators refused to allow the Coast Guard on board. All reports were sent to the Thirteenth District, which relayed the information to the Coast Guard cutter fleet. However, we saw almost no successful boardings as a result of that information. Many vessels in full compliance reported being boarded, sometimes repeatedly, during crab season with no violations.

The Coast Guard cutter fleet needs to do a better job searching for targeted vessels, rather than boarding the first vessel that crosses its path. Additionally, cutter fleet managers should do a better job of placing the cutters in positions where they will encounter crab boats during crab season.

Conclusions

The Coast Guard has a clear strategic mandate to eliminate deaths associated with commercial fishing. Previous efforts focused on voluntary safety examinations at the dock and enforcement during at-sea boardings. These approaches—dockside examinations and at-sea boardings—have had mixed success, due in the first instance to their voluntary nature and in the second to relative infrequency depending on the availability of expensive Coast Guard cutters. Both approaches share another critical flaw: high-risk fisheries are treated no differently than those with lower risk.

The Oregon and Washington Dungeness crab fishery is among the nation's most hazardous because of weather extremes, treacherous river bars, proximity to the surf, vessel instability, and intense market pressure leading to a "race for fish." Sixteen fishermen have lost their lives in the past seven crab seasons. A fatality rate per 100,000 workers was calculated for this fishery, adjusted for risk exposure, and found to be twice the national fatality rate for commercial fishing and comparable to the rate of deaths in the Bering Sea crab fisheries.

Given the success of the at-the-dock enforcement efforts started in 1999 in Alaska and the loss of six fishermen and three vessels off the Oregon and northern California coast that same winter, effective action by the Thirteenth Coast Guard District was clearly needed. The nature of the fishery, with its predictable locations and timing, easily lent itself to a specific, targeted at-the-dock operation. This was a just-in-time outreach to check stability and crab pot loading, EPIRBs, and life rafts on the vessels most at risk, but with a new dimension: Vessels with especially hazardous conditions would be prohibited from fishing until corrections were made.

The goals of Operation Safe Crab were simple: Examine as many crab vessels as possible, apply uniform standards across the fleet, focus on high-risk causal factors, keep exams short to minimize disruption to vessels, keep the Coast Guard examiners safe, provide credible consequences for noncompliance, involve the Coast Guard's law enforcement personnel, ensure that the operation was well publicized, and make the best use of Coast Guard resources. Thorough planning was the key to meeting these goals. The objectives and methods were communicated among the key players, and carefully selected examiner teams were deployed with data collection tools and reporting documents. Issues of Coast Guard authority were discussed, and clear rules of engagement were given to all examiners.

In the last 3 years, Operation Safe Crab has reached nearly 100% of the Oregon and Washington Dungeness crab fleet. Vessel discrepancy rates have steadily dropped. Critical lifesaving equipment has been brought into proper working condition. Although the numbers are small, a decrease in the number of deaths and vessel losses causing a death has been noted from the 3 years immediately prior to the first Operation Safe Crab and the 3 years since. Anecdotally, we believe safety awareness and spending by vessel owners on safety gear have increased.

A detailed discussion of the policies and actions of our last effort, Operation Safe Crab 2002, was presented. Several lessons learned were given, including the need for better communication and coordination with Coast Guard law enforcement personnel. In the future, continued safety improvements can be accomplished through reallocation of law enforcement effort from fishery management enforcement to fishing safety enforcement.

We believe that following our methods, other regional regulatory safety authorities can attain similar results with a cooperative, risk-based approach to prevention.

References

Medlicott C (2002). Using dockside enforcement to compel compliance and improve safety. *In* Proceedings of the International Fishing Industry Safety and Health Conference (Woods Hole, Massachusetts, Oct. 23-25, 2000), Lincoln JM, Hudson DS et al., eds. Cincinnati, OH: National Institute for Occupational Safety and Health. DHHS (NIOSH) Pub. No. 2002-147.

Perkins R (1995). Evaluation of an Alaskan marine safety training program. *Public Health Report, pp.* 701-702.

US Coast Guard, Fishing Vessel Casualty Task Force (1999). *Dying to Fish, Living to Fish.* Washington, DC.

Woodley C (2000). Developing regional strategies in fishing vessel safety: integrating fishing vessel safety and fishery resource management. Seattle: University of Washington.

Notes

a. The Thirteenth Coast Guard District comprises the states of Washington, Oregon, Idaho and Montana, and the adjacent ocean waters out to approximately 1200 nautical miles.

b. During the 2001-2002 season, one typical crabber was so overloaded with a tall pot load that the vessel rolled and capsized in the channel moments after the vessel left the dock. Apparently, the roll induced by the vessel's maneuvering (i.e., use of the rudder) exceeded the vessel's angle of positive righting energy, and she slowly rolled until the pilothouse hit the river bottom. No one was seriously injured.

c. Free-surface effect refers to the dangerous loss of stability when a liquid load is free to slosh transversely. The liquid's center of gravity moves as the surface seeks to remain level during the vessel's roll, resulting in a virtual rise in vertical center of gravity (and loss of stability).

d. Casualty data taken from official US Coast Guard records, as collected by the Fishing Vessel Safety Division (G-MOC-3) at Coast Guard Headquarters.

e. Due to data confidentiality concerns, a random vessel identifier replaced actual vessel ID numbers.

f. Data compiled by Christopher Woodley, 2000, during the writing of his master's thesis and e-mailed to the author.

g. Coincidentally, Captain James Spitzer had just been reassigned to Portland as the Captain of the Port after completing the Coast Guard's Fishing Vessel Casualty Task Force Report. He agreed wholeheartedly that more direct action was needed.

EVALUATION OF ALASKA'S COMMERCIAL FISHING VESSEL SAFETY PROGRAM

Alan J. Sorum, MPA
City of Valdez Small Boat Harbor
Valdez, Alaska, USA
E-mail: asorum@ci.valdez.ak.us

Executive Summary

Historically, commercial fishing is the most dangerous occupation in Alaska. Remote fishing grounds, poor weather, and cold water compound the problem. Loss of life prompted Congress to pass safety legislation in 1988. Alaska's Commercial Fishing Safety Program is a direct response to this legislation. The Coast Guard and other major organizations have developed a multifaceted program. Among its components are public education, law enforcement, and voluntary dockside fishing vessel safety examinations.

This evaluation involves three questions. First, are there unrealized opportunities or deficiencies present in the current fishing vessel safety program? Second, has the commercial fishing vessel safety program improved overall safety? Finally, does current, but unutilized, research exist that could help to improve fishing vessel safety? A three-part approach was designed to answer these questions.

1. A qualitative interview conducted with fishermen and representatives of the program.
2. A fishing vessel casualty and fatality data analysis .
3. A review of current literature and research.

Results show strong support for the dockside safety examination program. Deaths and vessel losses have declined in Alaska since the program began. Research suggests a measurable reduction in deaths, national acceptance of some Alaska safety practices, and the potential opportunity to predict specific dangers for Alaska's fishing fleet.

Recommendations developed in this evaluation suggest an increased focus on promoting dockside safety examinations and improving communications among participants. Fishermen value dockside examinations as an opportunity to ensure compliance with complex regulations. Coast Guard representatives favor them as an opportunity for a positive interaction with fishermen. Research shows those fishermen included in initial safety program planning and policy-making efforts continue with their involvement and participation in the safety programs. Many fishermen echoed this thought. They are much more comfortable with continued regulation when they can be part of the safety-related process.

Introduction

Commercial fishing is the most hazardous occupation in Alaska. Poor weather, small vessels, darkness, remote fishing grounds, and cold water only compound the dangers of fishing here in Alaska. Alaska's commercial fishing industry has lost an average of 34 vessels and 23 lives a year for the last 15 years (NIOSH 1997).

During the early 1990s, the death rate for fishermen in Alaska was 200 deaths per 100,000 workers per year (Conway, Lincoln et al. 1999). The overall rate for all workers statewide during the same period was almost five times the national average of seven deaths per 100,000 workers per year. The Coast Guard established the Commercial Fishing Vessel Safety Program (CFVS) to help address this tragic loss (US Coast Guard 1999).

Commercial Fishing Vessel Safety Program
The CFVS program is primarily a Coast Guard scheme to improve safety in the fishing industry. A major driver for establishing this program was the Commercial Fishing Industry Vessel Safety Act of 1988 (CFIVSA)(Williams 2000). The National Institute for Occupational Safety and Health (NIOSH), National Weather Service, Occupational Safety and Health Administration (OSHA), and the State of Alaska, Section of Epidemiology, are also considered major contributors to the Alaska CFVS program.

The current CFVS program is comprised of many different elements. Among these are public education, law enforcement, and voluntary dockside safety examination of fishing vessels. OSHA enforces federal standards for workplace safety, and NIOSH conducts research into improving safety and offers suggestions on how to implement it (NIOSH 1994, 1997; Lincoln and Conway 1999).

Elements of evaluation for the Commercial Fishing Vessel Safety Program

This evaluation of the CFVS program in Alaska addresses three separate questions:

1. Are there unrealized opportunities or deficiencies present in the current fishing vessel safety program?
2. Has the commercial fishing vessel safety program improved overall safety?
3. Does current research exist that could be used to improve fishing vessel safety?

This evaluation uses several approaches to answer these questions. The first is a qualitative interview with those involved in the industry, fishermen, and Coast Guard marine safety personnel. This effort seeks to find common thoughts on the program and hopes to identify unrealized opportunities. Second, a review of fishing-related fatality and vessel casualty data in Alaska is necessary to see if there are any apparent trends or relationships. Finally, this evaluation reviews fishing vessel safety research to gauge the effectiveness of the CFVS program. A research review can also identify missed opportunities and strategies.

Qualitative interview with those involved in the industry

Fishermen and the CFVS program sponsors were interviewed to see if any opportunities or deficiencies in the current program had been overlooked. The interviews provided a way to gauge the effectiveness of the program and its acceptance among those involved in it. This process also generated additional suggestions on how to improve the program.

Interview design

A six-question interview format was designed to gauge attitudes and perceptions of those involved in the CFVS program. The questions were open-ended and offered the opportunity for extended discussion of pertinent topics. The interview started from a general viewpoint and worked toward specific topics. Eight people actively involved in the CFVS program and another 10 people currently working as commercial fishermen were selected for interviews, but the constraints of time and access to those involved limited a truly random selection process. Group sampling appeared to be less important among the personnel working in the CFVS program, since the interviews covered almost all of those working within the Coast Guard's portion of the program in Alaska.

Data derived from the process are used in a number of ways: To identify safety topics developed through the interview process, to determine if there are common safety themes shared between the program sponsors and fishermen, and to develop descriptions showing the relationships existing between the program sponsors and fishermen.

Qualitative interview data and discussion of results

Sponsors and fishing personnel were asked to recall efforts made to improve fishing vessel safety. The most common responses from fishermen were with regard to the dockside safety examinations and safety equipment required as part of CFIVSA. Coast Guard personnel focused entirely on training efforts, including drills, stability, and damage control.

When asked to identify the most effective activity being conducted, both groups overwhelmingly selected the dockside safety examinations. An interesting point was that to most fishermen the dockside exams were important because they helped with regulatory compliance, while the program sponsor's personnel favored the exams as an opportunity to interact effectively with the fishing industry.

Opinions differed among fishermen questioned about what would be the least effective part of the CFVS program. Half of them indicated no problems with the program and were glad to have the Coast Guard available. The majority of Coast Guard personnel who responded focused on an overall lack of vessel inspections. They felt there was a failure to promote more dockside safety examinations. There was also support for more extensive periodic inspections of machinery and hulls.

The fourth question in the interview was designed to gauge the understanding both parties in the program might have of each other. Fishermen, in general, felt the Coast Guard often uses broad, general regulations and tries to apply them to unique, specific conditions. Many fishermen mentioned the need for a more realistic, rational approach to rule making. Regulations are easier to adopt than apply in the working world. Coast Guard personnel expressed a true interest in fishermen's safety and wanted fishermen to know this. A few respondents felt there could be lack of trust between the communities and that this needed to be addressed.

Both groups were asked to identify the one rule they would impose to improve fishing vessel safety. The majority opinion expressed by members of both groups was to enforce mandatory dockside safety examinations of fishing vessels. The consensus was to ensure compliance on vessels prior to their leaving the harbor.

An overall observation made after talking to these interested participants in the CFVS program is that both program sponsors and fishermen support the prevention of marine casualties. Both sides share more common attitudes than they might have initially believed. Saving lives and vessels can create strong bonds.

Table 1: Fatality rate information on Alaska commercial fishing industry

Year	No. of vessels lost	No. of persons on board	No. of persons killed	‡Fatality rate, %
1989	35	119	30	25
1990	31	137	31	23
1991	35	104	28	27
1992	46	114	33	29
1993	21	86	18	21
1994	36	266	13	5
1995	27	118	18	15
1996	39	116	25	22
1997	32	93	3	3
1998	33	145	13	9
1999	31	148	17	11
2000	19	97	8	8
2001	30	130	24	18
2002	22	114	12	10

Source: USCG Alaska District, Fishing Vessel Safety Coordinator.
‡Case Fatality Rate = (number killed / number at risk) x 100.
˚Data from Arctic Rose not included here.

Fishing-related fatality and vessel casualty data

The Alaska District of the Coast Guard provided information concerning all fishing vessel accidents since 1989 (Table 1). Sue Jorgenson, Coast Guard CFVS coordinator, was very helpful in describing the material. Coast Guard personnel in Valdez also contributed information specific to Prince William Sound and background on the voluntary dockside examination program.

NIOSH first illustrated the usefulness of calculating case fatality rates by year in the commercial fishing industry. We have reviewed the NIOSH process and updated the published information through 2002 and found the same results. Analysis of Coast Guard data shows vessel casualties have remained relatively constant, with a mean number of vessels lost equalling 31.2 and a median equalling 31.5 since 1989. The number of people on board and at risk has also been constant: a mean of 127.6 and a median of 117. The fatality rate is the relationship between those put at risk and the number killed in vessel casualties. This rate has improved since 1989.

One major problem with vessel statistics is the variability of losses. The commercial fishing fleet has a wide variety of vessel sizes and if a larger vessel carrying more people sinks, it can skew the totals. The sinking of the *Arctic Rose* greatly affected the fatality rate for 2001.

The Coast Guard data set classifies fishing vessel casualties by cause and includes capsizing, sinking, fire/explosion, towing, and grounding. Research suggests that capsizing and sinking incidents pose the greatest risk to the crew (Figure 1) (Lincoln and Conway 1995, 1999; NIOSH 1997; Conway, Lincoln et al. 1999).

Fishing-related deaths can also be classified by cause (Figures 2 and 3). "Man overboard" is the major cause of death for fishermen not involved with a vessel casualty. Sinking and capsizing comprise 50% of the deaths in incidents that involve a vessel casualty.

Data were also provided by the Coast Guard on the number of voluntary dockside safety examinations conducted since the beginning of the program. The Y-axis in Figure 4 was constructed with a logarithmic scale to allow examination data to overlay casualty information. Visually, the chart shows a slight decrease in vessels lost and an apparent improvement in the total number of lives lost. This observation would match well with the CFVS pro-

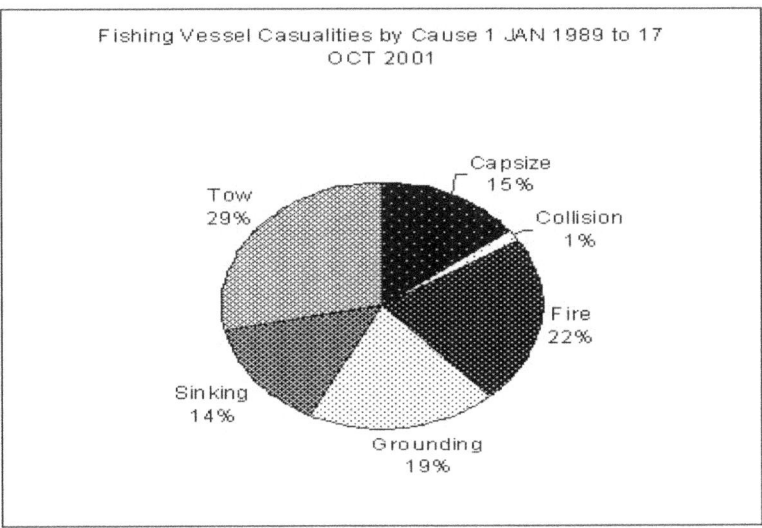

Figure 1: Fishing vessel casualties by cause, 1 January 1989 to 10 December 2002.
Data Source: USCG Alaska District, Fishing Vessel Safety Coordinator.

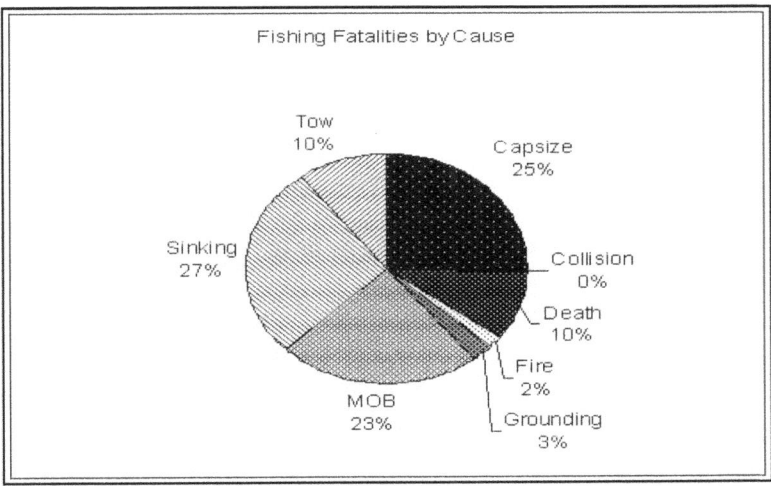

Figure 2: Fishing-related fatalities by cause,1 January 1989 to 10 December 2002.
Data Source: USCG Alaska District, Fishing Vessel Safety Coordinator.

gram emphasis on protecting lives after an accident has occurred. The other observation is that the total number of examinations given has remained very flat, but has decreased slightly since 1994.

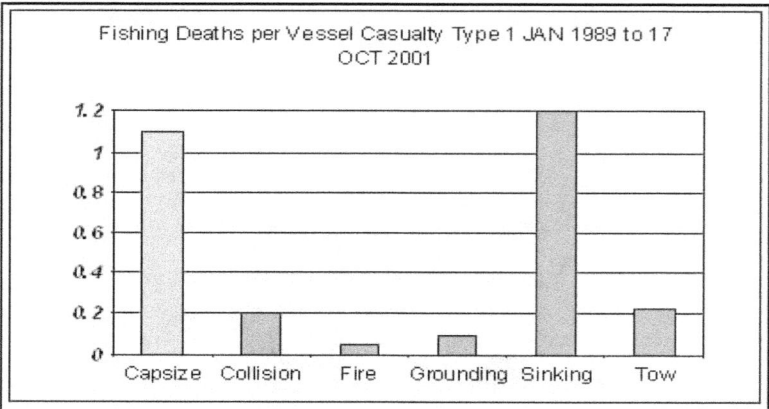

Figure 4: Fishing deaths per vessel casualty type, 1 January 1989 to 10 December 2002.
Data Source: USCG Alaska District, Fishing Vessel Safety Coordinator.

Research and literature review

Limited research is available on commercial fishing vessel safety. Related research and literature included in this evaluation came from searches of on-line resources, library records, Coast Guard references, and contacts with various agencies. NIOSH is the research leader in this field. Jennifer Lincoln, of the NIOSH Alaska Field Station in Anchorage, provided the research for this evaluation.

Numerous and lengthy lists of recommendations are available from government, education, industry, and trade groups. It is apparent that many of the same recommendations have been made repeatedly over the years. No one causal factor can improve fishing vessel safety because fishing vessel safety is a puzzle of competing interests and concerns. Rather than list other researchers' recommendations, this section of the evaluation will address information directly supportive of Alaska's effort or present information not noted at the national level.

Research conclusions can be broken into a number of related topics. These describe the effectiveness of Alaska's CFVS program or offer possible improvements to it. They include—

+ The Alaska CFVS program benefits from the specific mandates implemented from the CFIVSA.

+ Recent CFVS program efforts on a national scale have incorporated many of the safety concepts developed in Alaska.
+ The incidence of fishing-related fatalities and vessel casualties can be predicted through the use of statistical analysis. This analysis evaluates external factors that impact the fleet.
+ The commercial fishing industry has experienced a measurable reduction in fatalities and vessel casualty rates.

NIOSH findings regarding the Alaska CFVS program

CFIVSA mandated a number of new requirements for safety equipment and training on fishing vessels. These included the use of immersion (survival) suits, life rafts, and radio beacons. Required training includes first aid and emergency drills. Reductions in fishermen's deaths have mainly been the result of keeping them afloat and dry after they enter the water (Lincoln and Conway 1995, 1999; NIOSH 1997; Conway, Lincoln et al. 1999). Eighty-eight percent of all fishing deaths are caused by hypothermia and/or drowning.

Program mandates led to a significant decline in deaths from 1991 to 1998. One criticism of the program is that lifesaving efforts are all post-accident and that not enough is being done to prevent accidents.

National adoption of Alaska practices

Efforts made in developing the Alaska CFVS program have contributed to national efforts to improve fishing vessel safety. The Fishing Vessel Casualty Task Force was established in 1999 to address an alarming increase in fishing accidents on the East Coast. Eight NIOSH Alaska Field Station recommendations for improving fishing vessel safety were adopted by the task force (US Coast Guard 1999).

Prediction of fishing-related losses

Research shows that some fisheries are more dangerous than others (Lincoln and Conway 1995, 1999; NIOSH 1997; Conway, Lincoln et al. 1999). For example, the Bering Sea is especially dangerous because of the poor weather typical for its season. Vessels fish for a limited amount of stock and have only a narrow window in which to fish. Factors affecting safety include vessel size, market price of fish, and experience (Lincoln and Conway 1995, 1999; NIOSH 1997; Conway, Lincoln et al. 1999).

One group of researchers developed a statistical model that predicts which fishing conditions pose the greatest risk to those involved (Jin, Kite-Powell et al. 2001). Conclusions of the study suggest that capsizing and sinking accidents pose the greatest chance of resulting in the total loss of a vessel (Figure 4). An increase in the price of the fish harvested was associated with a decreased chance of a vessel loss. The risk of a fatality is greater in capsizing and fire and explosion accidents. Examination of the data describing Alaska's experience shows that sinking and capsizing events have the greatest risk for fishermen (Lincoln and Conway 1995, 1999; NIOSH 1997; Conway, Lincoln et al. 1999).

Future efforts in the CFVS program could utilize this research method to target and attempt to reduce high-risk situations. The methodology could accept details specific to Alaska and allow for a more customized interpretation of local data. The chart above shows a death per incident rate for various casualty types. It is apparent that accidents placing fishermen in the water are a major risk factor.

Recommendations

The Alaska effort to improve fishing vessel safety has been successful. That being said, no one involved in this industry believes commercial fishing is as safe as it could be. Recommendations developed from this evaluation focus on improving Alaska-based initiatives. There are many excellent recommendations being developed on the national level that have been previously suggested and may eventually be implemented here. In the meantime, actions developed, approved, and taken within Alaska will provide more immediate results.

The following recommendations were drawn from material developed by this evaluation, including analyses of casualty data, review of related research, and discussions with stakeholders. The list is not meant to be comprehensive, but is an attempt to generate discussion of common topics discovered in the evaluation process.

Involve fishermen
Swedish research points to the continued involvement and participation of those fishermen who were included in initial planning and policy-making efforts (Törner 2001). An outreach effort should be made by the Coast Guard any time new policies, procedures, or regulations are considered.

Coast Guard personnel felt that many fishermen believed the agency's role was to impose additional government regulations and increase the cost of doing business. Marine safety employees need to consider every contact with a fisherman as an opportunity to associate a human face with concern for safety.

One example of a positive contact is the Marine Industry Safety Day and Blessing of the Fleet sponsored by the Coast Guard's Valdez Marine Safety Office in 2001. Fishermen were able to practice the use of survival equipment and interact with Coast Guard personnel. The event was well received. It cast a positive light on the Coast Guard's safety efforts and was conducted at minimal cost. Other opportunities exist in the communities of Alaska. Health and safety fairs, high school career nights, and similar events are excellent avenues for public education.

Promote use of dockside safety examinations

Anecdotal comments support dockside examinations. What is apparent from the data is that the level of participation has remained flat since the program was developed. Coast Guard personnel and fishermen were almost unanimous in their support of the exams as an excellent method of interacting and educating fishermen about safety.

There needs to be further incentive for fishermen to participate in the program. The Ship Escort Response Vessel System (SERVS) provides oil spill prevention and response equipment for Prince William Sound. Fishing vessels under contract to SERVS to assist in spill cleanup are required to have a current dockside examination. Finding other organizations that could benefit from dockside examinations should be explored. These could include educational foundations contracting vessels for hire, agencies that place fisheries observers on board fishing vessels, and insurance carriers. Some of these efforts have already begun.

Wide support exists among program sponsors and fishermen for dockside examinations to be made mandatory. Three principal reasons were given. First, current regulations are difficult to interpret. A second opinion helps with compliance. It pays to have outside inspections of required safety equipment. Second, most vessels involved with the program have been part of it for quite a while. Eighty percent of the vessels in Alaska do not participate. Third, the examination process offers a great opportunity for interaction among those involved that is not conducted under threat of punishment.

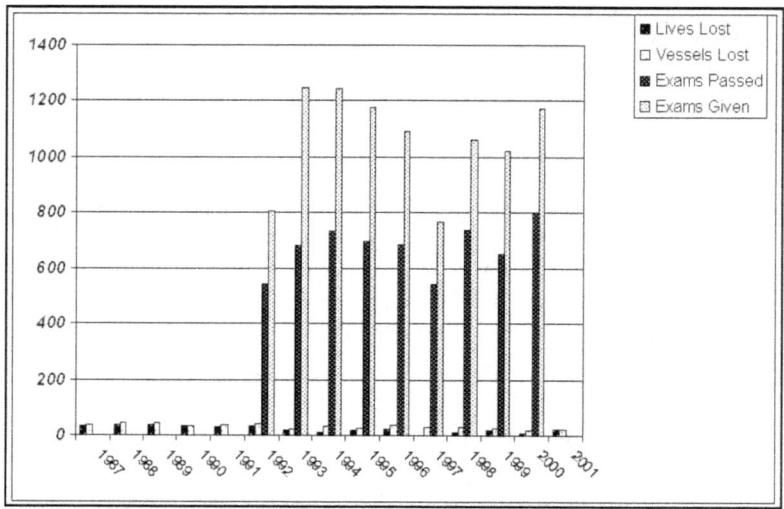

Figure 5: Fishing vessel casualities versus dockside examinations conducted.
Data Source: USCG Alaska District, Fishing Vessel Safety Coordinator.

Dockside examinations are not currently meant to be a strict compliance tool of the Coast Guard.

Conclusions

The number of deaths has decreased since the inception of Alaska's CFVS program (Figure 4). There is no doubt expressed by those interviewed from the fishing industry that there are still too many deaths and vessels lost each year. The Coast Guard has developed the framework of a successful program. Interviews with the program sponsors and fishermen show wide support for the CFVS program. These communities are more united in their concern for safety than either may realize. There are many valid recommendations in existence that could improve the CFVS program. What the Coast Guard lacks now are the tools to complete the mission and implement these important ideas.

References

Conway GA, Lincoln JM et al. (1999). Alaska's model program for surveillance and prevention of occupational injury deaths. *Public Health Report* 114(6): 550-558.

Jin D, Kite-Powell H et al. 2001. The safety of commercial fishing: determinants of vessel total losses and injuires. *Journal of Safety Research* 32(2): 209-228.

Lincoln JM and Conway GA (1995). Preventing deaths in Alaska's fishing industry. *Public Health Report* 110: 700.

Lincoln JM and Conway GA (1999). Preventing commercial fishing deaths in Alaska. *Occupational and Environmental Medicine* 56(10): 691-5.

Törner Marianne (2001). Finding way to support safety in the fishing industry. Proceedings of the Marine Safety Council, US Coast Guard, April 2001.

National Institute for Occupational Safety and Health (1994). Preventing drownings of commercial fishermen. DHHS (NIOSH) Pub. No. 94-107.

Lincoln JM and Conway GA (1997). Commercial fishing fatalities in Alaska: Risk factors and prevention strategies. Cincinnati, OH: National Institute for Occupational Safety and Health. Current Intelligence Bulletin 58. DHHS (NIOSH) Pub. No. 97-163.

US Coast Guard, Fishing Vessel Casualty Task Force (1999). *Dying to Fish, Living to Fish*. Washington, DC.

Williams J (2001). Fishing vessel safety action plan evaluation. *Proceedings of the Marine Safety Council* 58(2):60-63.

Appendix A: CFVS program interview questions

1. What is your involvement with commercial fishing?
2. What efforts can you recall being made to improve commercial fishing vessel safety?
3. The US Coast Guard has a commercial fishing vessel safety program in place. It is composed of many different components such as dockside examinations, research, regulation, investigation, and law enforcement. Is there part of this program you believe has been very effective? Why?
4. Is there part of this program you believe can be improved? Why?
5. The Fishing Vessel Safety Program involves people from within fishing and government communities. What do you think people from the other side of the program fail to realize about your community?

6. If you could propose one rule to improve fishing vessel safety, what would it be? Why?
7. Do you have any further comments that should be considered in the discussion of fishing vessel safety?

Appendix B: Selected respondent comments

- Stability testing of vessels should be required. Many fishermen don't realize how easy it is to roll a loaded vessel.
- Complete a damage control class. The knowledge and skills turned out to be lifesavers for fishermen facing a sinking vessel.
- Mandatory licenses for fishing vessel operators would insure knowledge and practical understanding of marine operations. Many fishermen have little or no experience operating fishing vessels.
- Emergency equipment required by the CFIVSA of 1988 is very important and only represents a minimal level of protection.
- Voluntary dockside safety examinations are very important. They help ensure equipment is functional, that the vessel meets current regulations and offers a great change for interaction between the Coast Guard and fishermen.
- The results of the marine casualty investigations should be readily available to the industry. The Coast Guard's "Lessons Learned" program is very useful.
- Dockside examinations of fishing vessels should be mandatory. Eighty percent of the fishing vessels in Alaska do not participate in a proven program that could save their lives.
- The Coast Guard needs to realize fishermen are operating small businesses. Inspections, regulations and boarding should be planned not to disrupt work whenever possible.
- Safety regulations are easier to draft than enforce. There needs to be a practical application of regulations impacting the fleet.
- There is no inspection required for fishing vessel rigging. Every year several people are killed by falling booms, broken pins, and winches. Deck safety need to be addressed by the industry and regulators.
- Vessel orientation and safety drills are very important. Classroom instruction is not as valuable as on-the-job training in the field. Documentation standards for on-board drills need to be strengthened.
- The Coast Guard is really doing a great job.
- Stress marine firefighting skills and equipment use. A vessel on fire at sea represents a major disaster for the crew.

THE SOCIETY OF NAVAL ARCHITECTS AND MARINE ENGINEERS INTEGRATED STABILITY TRAINING PROGRAM FOR COMMERCIAL FISHING VESSEL CREWS AND THE FISHING COMMUNITY

John E. Womack, Naval Architect, MSE, BSE
Mid-Atlantic Shipwrights
Pittsville, Maryland, USA
SNANE, Chairman of Working Group B, Ad Hoc Panel 12-
    Fishing Vessel Operations and Safety
E-mail: shipsjw@aol.com

Bruce Johnson, Professor Emeritus, PhD, MSME, BSME, USNA
Annapolis, Maryland, USA
SNAME, Chairman of Working Group A, Ad Hoc Panel 12-
    Fishing Vessel Operations and Safety
E-mail: aronj@bellatlantic.net

Summary

The commercial sector of the fishing industry has more diverse types and classes of vessels and fishing methods than any other category in the fishing industry. For this reason, the traditional apprenticeships from deckhand to captain no longer serve as an effective stability training method for fishermen. Under the Society of Naval Architects and Marine Engineers (SNAME) Ad Hoc Panel 12, Fishing Vessel Operations and Safety, a three-component, integrated stability training program using verbal and visual presentations, a written booklet, and a set of table-top hands-on demonstration models has been developed to address this issue. This program has been designed to be flexible for use on board individual vessels, as well as during large meetings where more technical presentation methods can be offered.

Brief history of the evolution of stability training

Until recently, the traditional method of training mariners in ship stability involved long apprenticeships for everyone on board, from deckhands to captains. Mariners-to-be would learn the proper "feel" of sailing the ships of the day under the tutelage of experienced captains. In spite of the crudeness of this approach, it was actually quite successful, and relatively few vessels sank solely because of stability issues. The stability of commercial sailing vessels was capable of being judged by feel because of the way they were propelled and their basic design characteristics.

In comparison, traditional hands-on apprenticeships for stability training on today's fishing vessels are a dangerous proposition. Many different configurations have been developed for fishing vessels to reflect various fishing techniques and local conditions. A subjective judgment of stability in vessels of one fishery cannot be transferred reliably to another fishery having different vessel types and gear. The different fishery vessel configurations have become, in general, more sensitive to stability problems related to shrinking freeboards, numerous openings in the watertight envelope, large and variable capsizing forces related to fishing gear, and significant changes in stability characteristics during the voyage. All of this creates a continually moving target of the vessel's feel that the crew cannot reliably predict.

This lack of predictability has led to the creation and subsequent transfer from captain to captain of many myths and misconceptions about how stability works and whether the feel of the vessel can be used to judge whether stability levels are adequate. The stability characteristics of modern commercial fishing vessels are very complex and can mislead crews, giving them a false impression that all is well when in actuality they may be in imminent danger of capsizing. These misconceptions continue to persist, not through the fault of the fishermen, but through the fault of naval architects, safety trainers, and regulators who have failed to introduce stability training adequately to crews. (For additional discussions on this topic, see Appendix A: Evolution of Stability Training.)

Requirements of an effective stability training program

To rectify problems associated with stability training and improve crew safety, effective means are needed to teach stability to the crews. A few programs are in limited use today, but they tend to have one or more common failings that prevents their practical use. To be effective, a stability training program for the commercial fishing community needs to address numerous issues.

The first issue is for trainers to determine their target audience. While it is obvious that crews are the primary focus of stability training, any useful program must also target vessel owners, the Coast Guard (US Coast Guard 1999) and other inspection agencies, safety trainers, surveyors, insurance companies, and fishing regulators. Other parties that must understand stability issues include the International Maritime Organization (International Maritime Organization 1995; Francescutto 2002), the Food and Agriculture Organization (Turner and Petursdottir 2002), the International Labor Office (Wagner 2002), and accident investigation agencies (Lang 2002). The key to a successful training program is that all members of the commercial fishing community gain knowledge of stability. Vessel owners and their crews need to fully understand vessel stability so that they are able to emphasize and practice safe operations while maximizing productivity. More knowledgeable inspectors and surveyors will focus attention on vessel design and use that affect stability. Fishing regulators can better consider all aspects of a fisheries management plan, including how the fishing fleet can maximize safety, with a better knowledge of stability issues.

The second issue that must be addressed is flexibility. Given the wide audience desired, a successful training program needs to be flexible to adapt to various audience sizes, audience concerns, and presentation venues. Audience size will vary from a few participants, if addressing a single crew, to possibly a hundred, if a presentation is being made at a trade show. Likewise, appropriate venues will range from individual fishing vessels to large trade shows or fisheries management councils. A flexible training program must also be adaptable to handle the specific concerns of a single vessel or fishery, as well as the generic concerns of a large group.

The third issue is the need for hands-on learning. Book work alone cannot effectively deliver the lessons desired (Herbert 2002). By creating a means for crew members to experience a more hands-on lesson, skills will be transferred to the real-world workplace and make the training more believable and understandable.

The fourth issue is integration. A successful stability training program should be integrated throughout fishing vessel safety training programs. In addition, consistency in how the material is presented is necessary for better understanding of the topic.

The fifth issue is presentation. A common refrain in sales is "presentation is everything," and this is certainly true in stability training. To help "sell" the lessons, an effective presentation scheme must be used. This includes the primary use of figures and graphics over words. Those figures and the fishing operations they depict should represent real-world fishing vessels to enhance the transfer of skills and knowledge. In addition, videos of model tests and animations should be used to place the lessons into a real-world context, as well as provide a more entertaining learning experience. To aid visual impact, color should be used wherever possible. For example, by using red, yellow, and green for different stability curves, the relative safety level of each curve can be intuitively and quickly signaled to the audience.

The sixth issue is portability. Any presentation program needs to include learning aids that are easily transported to various venues. Going to crews, rather than having them come to the instructor, can increase the number of audiences. This portability requirement includes hands-on demonstrations.

The last, but perhaps most important, issue for an effective stability training program is the scope of the subject matter to be taught. The program should focus only on teaching the basic concepts of stability and how fishing operations affect a vessel's relative stability levels. No attempt should be made to teach how to calculate a vessel's actual stability. That is the province of a naval architect who fully understands all the complexities and nuances, many of which can contain hidden dangers.

While this simplistic approach may at first appear to fail to provide useful training, it will actually provide more practical knowledge to the maximum possible audience for several reasons. First, complex concepts require long commitments of time to be taught effectively, something most members of the fishing community don't have. It takes a minimum of 4 years of schooling at an institution of higher learning and many years of practical experience for a naval architect to fully understand stability.

Second, the educational level of fishing boat crews varies greatly. Advanced stability concepts are highly technical and rely on fully understanding "imaginary" concepts such as metacentric height. Third, the primary goals of a successful stability training program are to (1) demystify the stability guidance creation process, including the inclining test; (2) provide sufficient understanding of stabiity to allow crews to have faith in the stability guid-ance given even when it runs counter to past experiences or beliefs, and (3) provide a basic understanding of the way common fishing situations affect stability to allow crews to evaluate their risks in nonstandard loading, operat-ing, and weather conditions.

Thus, a focus on the basics of stability and a focus on preserving lives and vessels are what is needed for most commercial fishermen, not advanced engineering formulae.

Results

Description of the SNAME-developed commercial fishing vessel stability training program

The Society of Naval Architects and Marine Engineers (SNAME), under Ad Hoc Panel 12, Fishing Vessel Operations and Safety, developed a com-mercial fishing vessel stability training program that addresses the issues raised above. The training program consists of three principal components: a written booklet, a verbal and visual presentation, and a set of hands-on training models. All the figures presented below are examples of those used in the booklet, the presentation, or the hands-on demonstrations. In the booklet, these figures are three to four times larger than shown here to pro-vide good readability.

This stability training program has been under development for approxi-mately 3 years (Johnson, Wallace, and Savage 2002; Johnson and Womack 2001; Womack 2002). During this period, many concepts were developed and tested over a wide range of audiences. Three drafts of the booklet were made available for review on Ad Hoc Panel 12's Intranet site and also di-rectly sent to selected Coast Guard personnel, naval architecture firms, and safety trainers for their comments. In addition, a conceptual version of the hands-on demonstration model, including plastic tanks, was built for trial purposes (see Figures 1 and 2). This conceptual model and several draft ver-sions of the presentation were tested with five fishing vessel crews and own-ers when stability letters were presented, as well as with the Coast Guard's Commercial Fishing Vessel Industry Advisory Committee, several Coast Guard fishing vessel safety coordinators, members of the Maritime Lawyers' Association's Fisheries Committee, SNAME's Panel O-44 (Marine Safety and Environmental Protection), and various naval architects and shipbuild-ers both within and outside the fishing community.

Figure 1: Model, quarter view   Figure 2: Model, end view

Basic structure of the SNAME stability training program

As mentioned, the stability training program consists of three principal components. In this setup, the presentation and its accompanying set of demonstration models are intended to be the primary teaching tool. The booklet, while of use as self-teaching material by many readers, is intended as a take-home refresher manual. This approach is preferred because no matter how well-written the booklet, within the wider target audience, there will always be questions or topics that cannot be fully addressed during training sessions.

The main presentation consists of approximately 48 Power Point slides, although this number can vary to address individual audiences. Eight companion slide shows of about 14 slides each, which graphically animate the interactions in the hands-on demonstration models, are also available. Furthermore, a series of short videos demonstrating the dangers of operating in heavy seas can be used during the main presentation. The presentation generally takes between 1 to 2 hours, depending on how much of the material is used. When used for individual crews, the time required will generally be closer to 1 hour, which is both practical and tolerable.

The accompanying 8-1/2- by 11-inch (215- by 280-mm) booklet is approximately 60 pages long. In addition to the stability training material described in Section 3.2, the booklet also contains a glossary of key terms and a list of contacts for additional information. For the advanced reader, an appendix shows some of the methods naval architects use to gauge a fishing vessel's stability.

The last component of the stability training program is the hands-on demonstration models. Two sets of models have been built for table-top use at different venues. A large version with a hull length of about 24 inches (600 mm) is intended to be used in large meeting rooms. With all the accessory parts, this version requires two large (30- by 22- by 10-inch)(760- by 560- by 250-mm) and two small (26- by 17- by 9-inch)(660- by 430- by 230-mm) suitcases to transport. A half-scale version with a hull length of about 12 inches (300 mm) is also being developed for individual vessels. This set requires only one large suitcase to transport, making it feasible to use when giving a stability letter to a crew or at a small gathering.

Topics covered by the presentation and booklet

The booklet, presentation, and demonstration models cover five basic topics.

1. A description of stability and how it is created.
2. An explanation of how a vessel's stability can be graphically displayed.
3. Initial and overall stability conflict.
4. Relative effects of fishing and vessel operations on stability.
5. General stability and seamanship guidance.

The first section starts out by defining the differences between "stable" and "unstable" fishing vessels, as this is key to any further discussions. A stable fishing vessel is one that has sufficient righting forces available to counter all capsizing forces encountered during the fishing trip. An unstable vessel does not have sufficient righting forces. Next, the two forces of gravity (G) and buoyancy (B) acting on a fishing vessel to develop these righting forces (i.e., its stability) are introduced. The first section then explains how the interaction of these forces, specifically the shifting of buoyancy as the fishing vessel heels over, either creates a positive righting force or a negative capsizing force (Figure 3).

The second section follows by showing how naval architects can graphically represent the righting forces by using a righting arm curve (Figure 4). Using this curve, the stability characteristics of a vessel under different loading conditions can be evaluated and compared. This section also gives a brief explanation of the inclining test and how a vessel's center of gravity (G) and center of buoyancy (B) are calculated. The purpose is to demystify the stability guidance development process, which can, quite frankly, appear as black magic to crews. To them, the naval architect moves some weights back and forth on the deck, disappears, and then reappears with stability guidance.

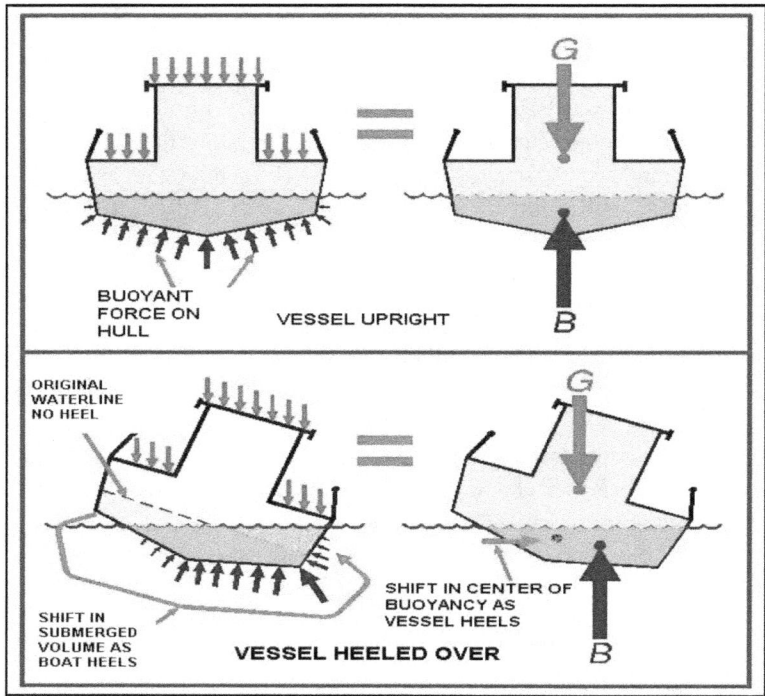

Figure 3: Shifting boyancy (B)

As noted, this sort of maneuver can sometimes run counter to the crew's beliefs.

The third section of the training program explains facts about initial stability versus overall stability conflicts, which is the root cause for many of the fishing community's misconceptions about stability. Quite often, a vessel's stability does not correlate with a crew member's perception of vessel stability based on the vessel's feel. As shown in Figure 5, the difference in stability levels between two loading conditions at high heel angles is quite significant. The problem is that the difference in stability levels is relatively small at the low heel angles typically experienced by the crews. It is from these low heel angles that the vessel's feel is derived by the crews, but, as shown this situation, feel cannot always provide a good indicator of the stability available during severe conditions.

The fourth section of the booklet and presentation shows the relative effect of typical situations on a vessel's stability. This section is broken into four

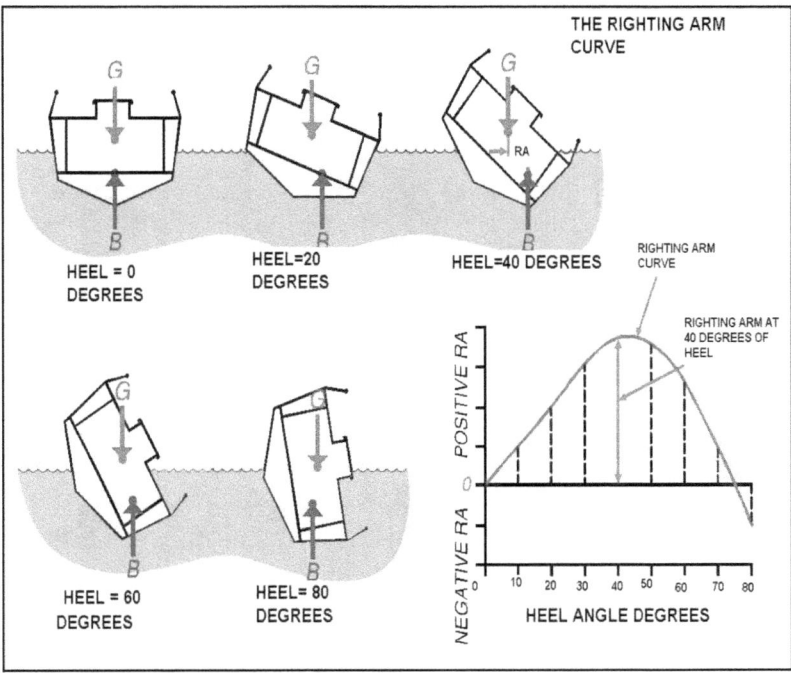

Figure 4: Righting arm curve

main topics: "Initial Versus Overall Stability," "Free Surface," "Fishing Operations," and "Vessel Operations." "Initial Versus Overall Stability" covers vessel loading situations such as overloading (Figure 6), adding ballast, and the cumulative effect of weight creep. "Free Surface" includes the effects of slack tanks, progressive flooding, water on deck, and water trapped in large deckhouses (Figure 7). "Fishing Operations" covers such topics as lifting weights, towing trawls (Figure 8), and shifting loads. The last topic, "Vessel Operations," covers vessel handling in heavy following (Figure 9), quartering, beam seas, icing (Figure 10), and the effects of beam winds.

The relative effects of a particular fishing situation on a vessel's stability levels are shown by way of the righting arm curves. This approach clearly shows the significant reduction in overall stability that can occur during typical fishing situations. The loss in stability is generally twofold: an overall reduction in the righting arms (righting forces) and a reduction in the outer range of positive righting forces. In some cases, such as when water is trapped in the deckhouse (Figure 7), the reduction in stability is threefold: an overall reduction in the righting arms (righting forces) and a reduction in both the inner and outer range of positive righting forces. However, as noted previ-

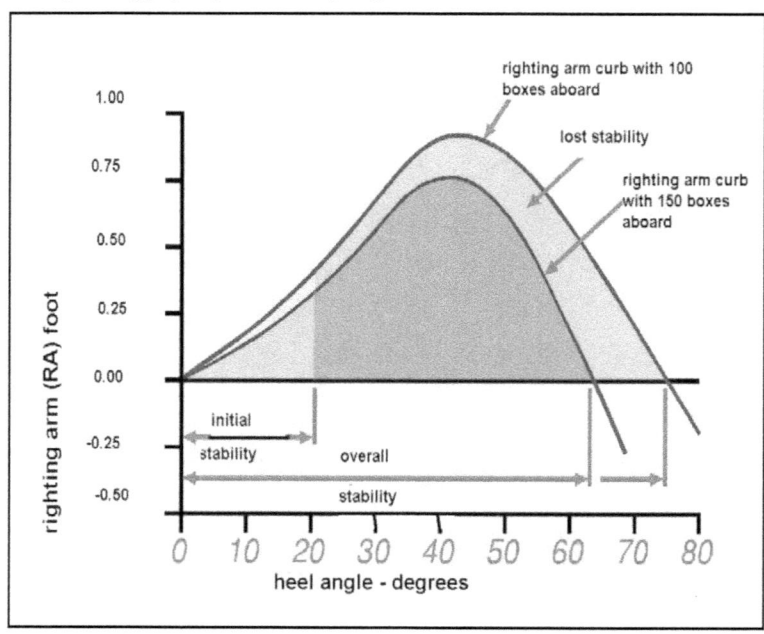

Figure 5: Initial versus overall stability

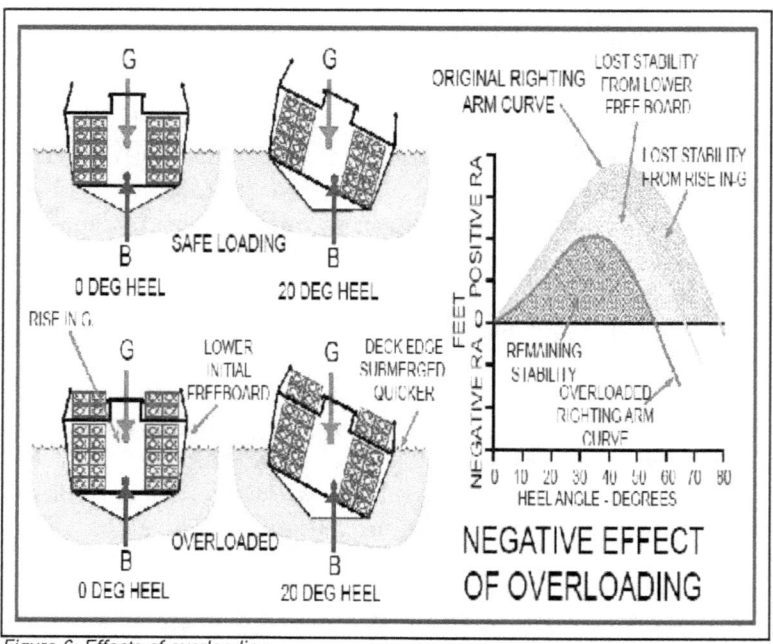

Figure 6: Effects of overloading

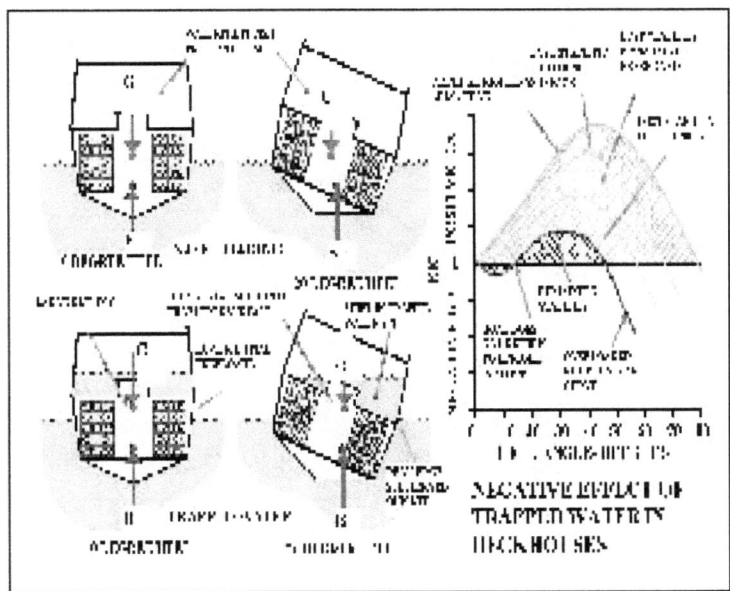

Figure 7: Effects of water

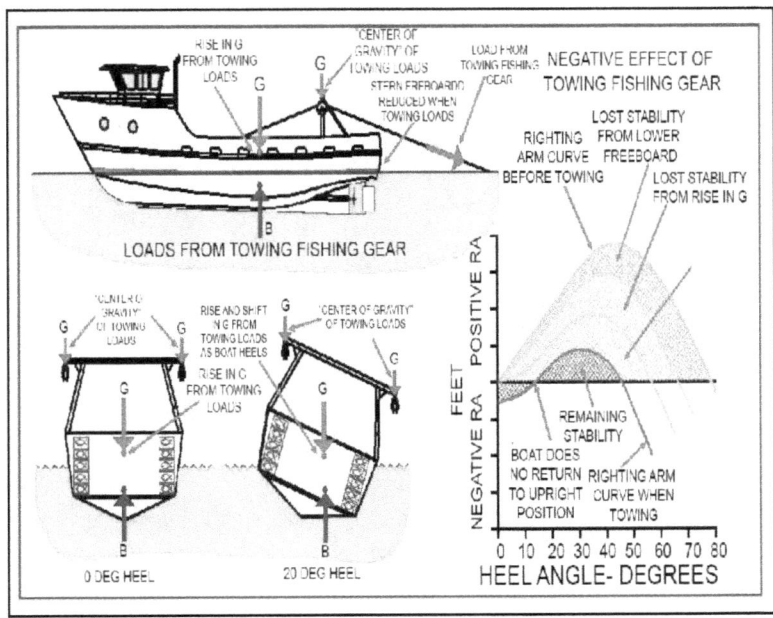

Figure 8: Effects of towing gear

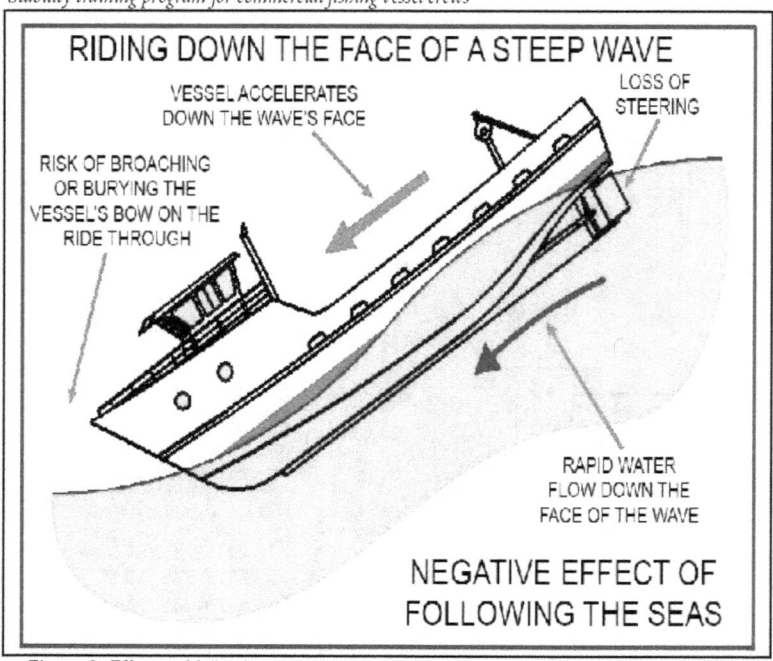

Figure 9: Effects of following seas

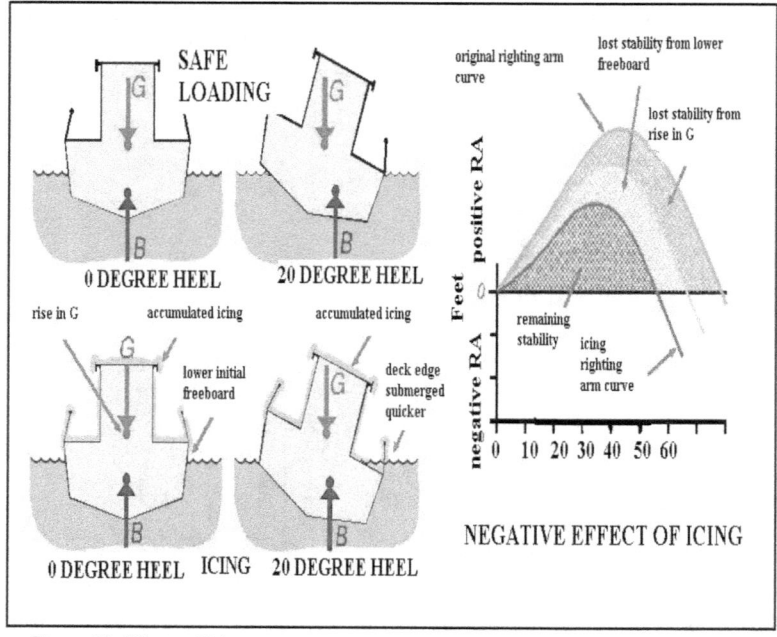

Figure 10: Effects of icing

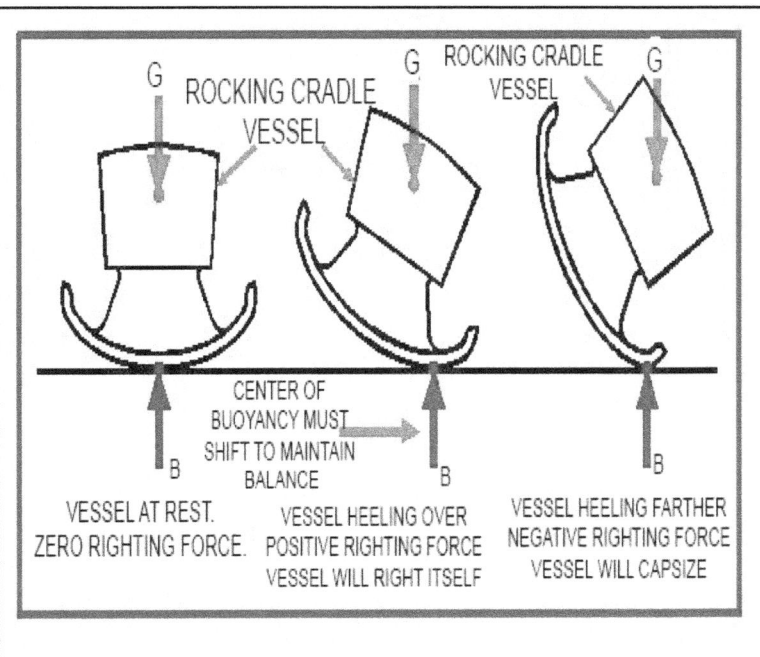

Figure 11: Baby's rocking cradle

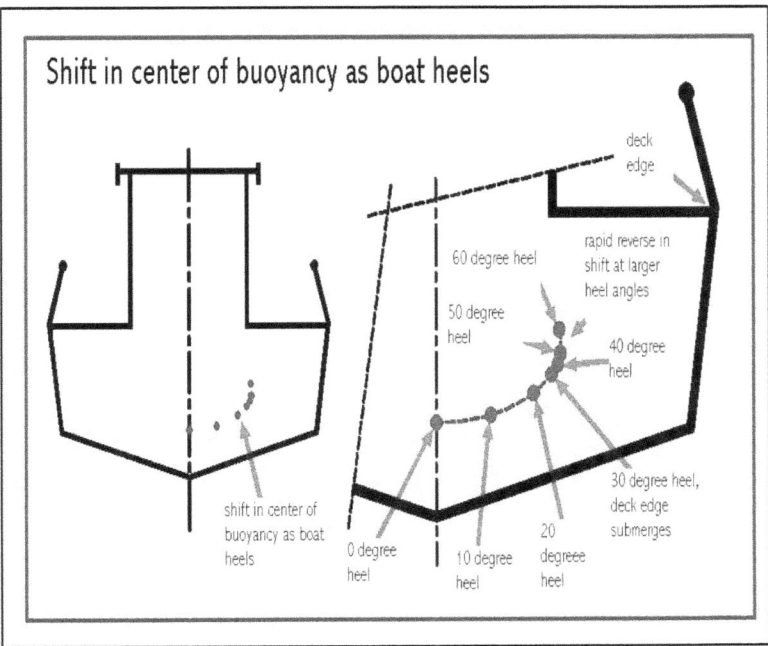

Figure 12: Center of buoyancy path

Figure 13: Model at even keel

Figure 14: Model heeled over

ously, the impact on initial stability levels, and thus the feel of the vessel, is often minimal and does not correctly convey the magnitude of the reduction in stability levels that has occurred.

This reduction is shown visually by indicating the initial or safe righting arm curve in green and the final or unsafe curve in red. Because many environmental conditions (Figure 10) contribute to a vessel's instability, yellow righting arm curves demonstrating individual impacts are also shown. This exercise further reinforces the take-home message that stability levels can decrease very quickly from multiple causes without the crew being aware of the dangers present. In the example of icing (Figure 10), stability loss occurs from the combined impacts of the added weight of the ice high on the vessel and the loss of freeboard.

Figure 15: End view

Figure 16: Demonstration model, stern quarter view

The last section of the training program provides general guidance on stability and seamanship to assist crews in preserving their vessel's stability. The topics covered include maintaining watertight integrity, developing and following stability guidance, and prudent vessel operations. These are generic lessons applicable to most commercial fishing vessels. They are also intended to encourage additional discussions by the audience if there are any of the guidance suggestions audience members are uncertain about.

Description of hands-on demonstration models
The hands-on demonstration models are the third and final component in the SNAME-developed stability training program. As illustrated by the material in the booklet and the presentation, one of the key teaching concepts is to equate the ability of a fishing vessel to stay upright with the ability

Figure 17: Interchangeable rockers

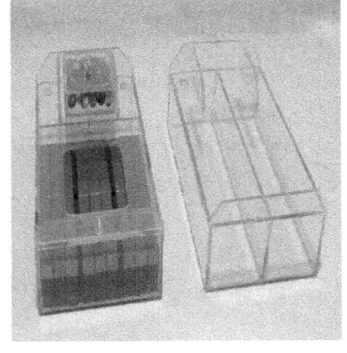

Figure 18: Typical hull inset

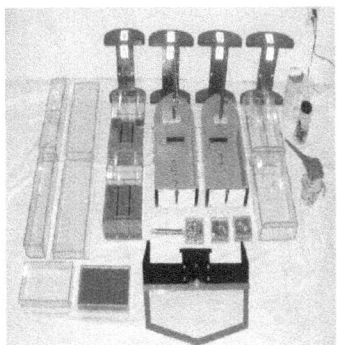

Figure 19: Set sample

of a baby's rocking cradle to remain upright. Both the vessel and the cradle actually work the same way; the point where the cradle's rocker touches the floor is the same as a fishing vessel's center of buoyancy. As shown in Figure 11, as long as the cradle's "center of buoyancy" shifts faster outboard than the cradle's center of gravity, the cradle (and baby) will happily return upright.

What makes this cradle analogy effective is the fact that the typical path of a fishing vessel's shift in center of buoyancy as the vessel heels over, as shown in Figure 12, is very similar to that of a cradle's rocker. By shaping the model's rocker to match the fishing vessel's center-of-buoyancy path, the model will correctly replicate the intractions between a fishing vessel's center of gravity and its center of buoyancy. Figures 13 and 14 show the hull's amidships section overlain on the demonstration model. The blue portion is the model's rocker. Note the black arrows, labeled B, that show the rocker's contact point moving at the geometric center of the submerged hull shape as the model heels. This allows relative changes in the model's initial stability and overall stability, its capsize angle, and variations in righting forces as the model heels to be modeled correctly. These interactions allow the audience to experience directly the effects of common fishing situations safely at high angles of heel.

The base model setup consists of a "hull" (Figures 15 and 16) and a rocker base to which a series of auxiliary components are attached in order to model a series of common fishing situations. Three different detachable rocker bases (Figure 17) are used to model changes in the center of buoyancy's path with changes in the vessel's displacement accurately. From this base model setup, sets of tanks, cargo holds (Figure 18), and/or a removable mast are added for the desired demonstration. The components described below

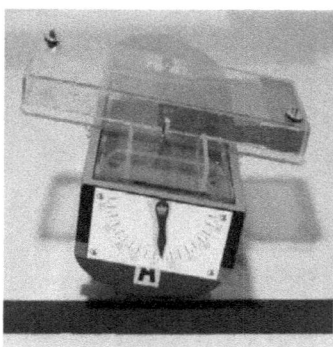

Figure 20: Water on deck setup

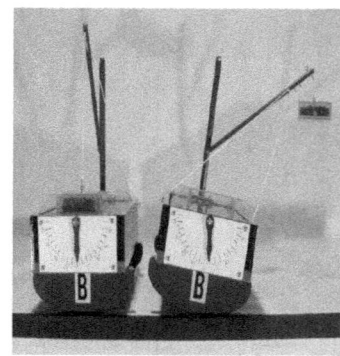

Figure 21: Lifting weights setup

contain sufficient parts to allow for two model setups to be run side by side for direct comparison, although a one-hull model setup produces acceptable demonstrations.

Currently the following principal components have been developed (Figure 19):

> Two hulls with heel angle indicator.
> Four detachable rocker bases—one type A, two type B, and one type C.
> Two removable mainmasts with lifting booms.
> Two hold free surface tanks, no centerline bulkhead.
> One deck free surface tank, no centerline bulkhead.
> One deck free surface tank, with centerline bulkhead.
> Two hold dry cargo boxes with two sets of cargo weights.
> Two deck dry cargo boxes with one set of cargo weights.
> Two lifting weight boxes.

Using these components, the relative effects on a fishing vessel's stability can be demonstrated. These will cover most of the typical fishing situations encountered by crews.

> Progressive hull flooding: dry, 10%, and 25% flooding.
> Water on deck: dry, 15%, and 30% flooding (Figure 20).
> Water trapped in a wide deckhouse: dry, 15%, and 30% flooding.
> Overloading.
> Overloading with hull flooding.
> Lifting weights on centerline.
> Lifting weights over the side (Figure 21).
> Lifting weights over the side with hold free surface.
> Effects of adding a centerline bulkhead.

Descriptions of video and animation slide shows

The presentation can be further enhanced by the inclusion of companion videos and animations of the hands-on demonstration models. These additions make the learning process more interesting, bring the lessons into the real world, and reinforce the concepts discussed regarding the interactions of buoyancy and gravity in the hands-on demonstration models.

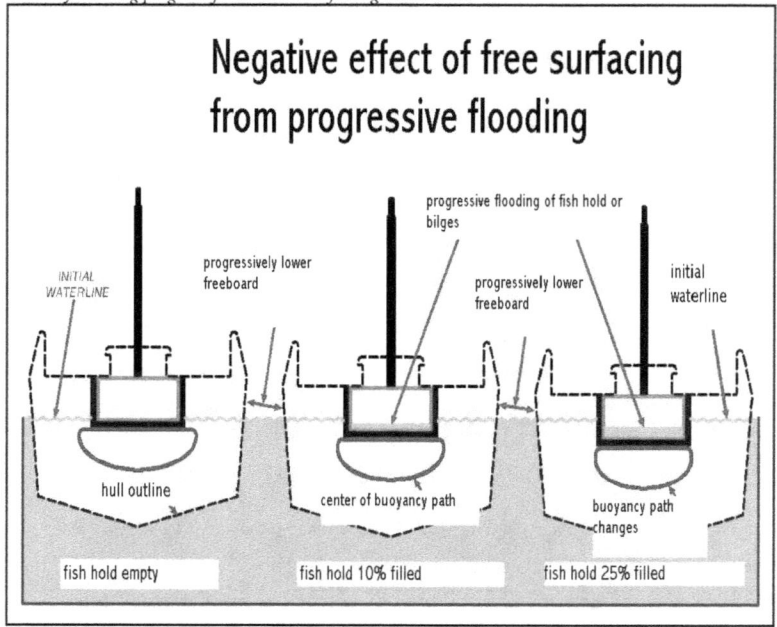

Figure 22: Example of animation for progressive flooding into vessel's hull

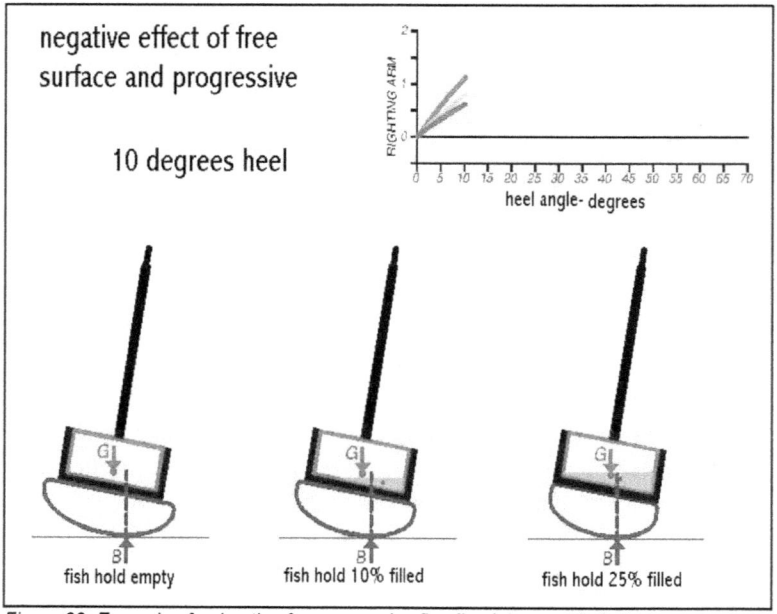

Figure 23: Example of animation for progressive flooding into vessel's hull

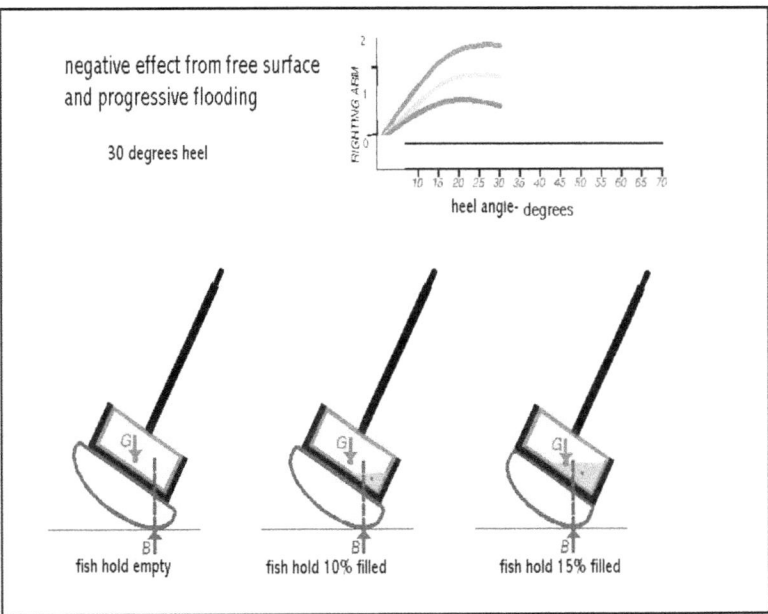

Figure 24: Example of animation for progressive flooding into vessel's hull

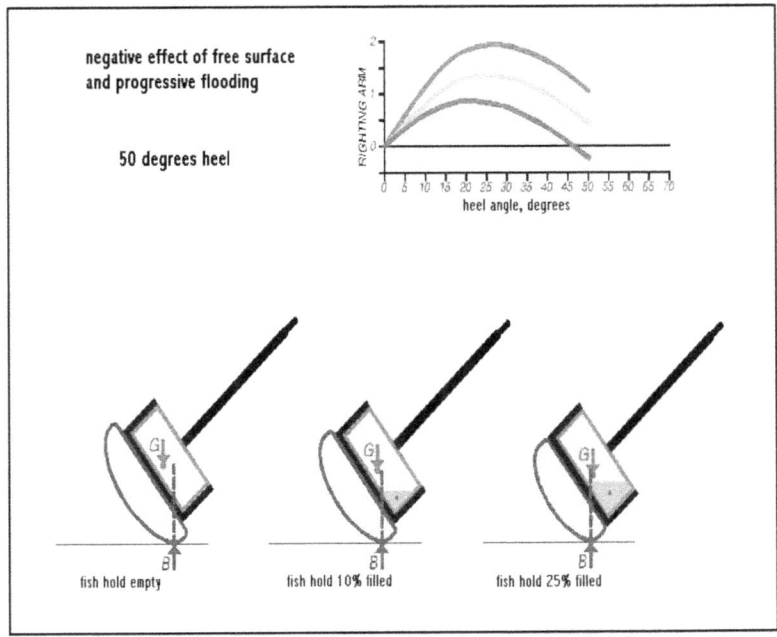

Figure 25: Example of animation for progressive flooding into vessel's hull

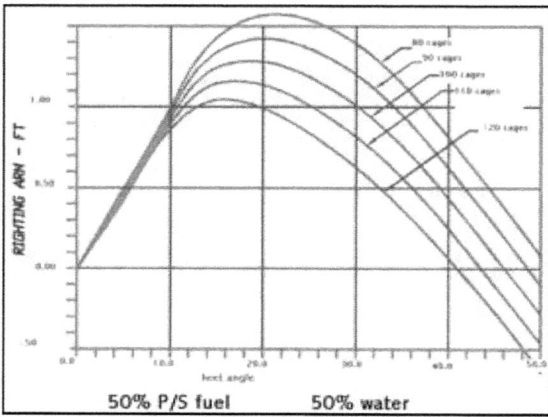

Figure 26: Vessel's righting arm curve

The current videos consist primarily of fishing vessel model tests. These all use free-running models operating in regular and irregular waves. The model tests in regular waves are from the National Research Institute of Fisheries Engineering of Japan (Umeda, Matsuda et al. 1999) and show broaching, loss of stability on a wave crest, and bow diving. The model tests in irregular waves are from a study by the Institute for Marine Dynamics, National Research Council of Canada (Grochowalski 1989). These videos show the full range of fishing vessel operations, including shipping water on deck, broaching, loss of stability on a wave crest, and bulwark tripping. Two videos of real fishing vessels operating in heavy seas are also available. The first shows a fishing vessel being capsized by a stern quarter-breaking wave, and the second shows a fishing vessel bow diving into an unexpectedly large wave face, blowing out the pilot house windows.

The animations of the hands-on demonstration models are used to show the interactions of the center of gravity and the center of buoyancy as the models are heeled over. These animations are done in 5° (0.087-radian) increments. Two or three side-by-side examples are used to show the relative effects on stability directly. Figures 22, 23, 24, and 25 are examples of the animation showing progressive flooding into the vessel's hull. These animation slide shows are particularly effective by having the righting curves, in color, develop in motion as the model heels over.

Flexible approach of SNAME stability training program

The SNAME-developed stability training program uses the Microsoft Pow-

tation medium. This allows the presenter to modify the training program easily to suit the particular needs of the audience. For a specific fishery or a specific boat, potentially confusing parts of the presentation not applicable to the subject audience can be removed. For example, if the vessel or fishery in question does not tow fishing gear, then those slides dealing with towing gear can be quickly removed. And conversely, the presenter can also easily add slides to reinforce topics critical for the target vessel or fishery.

Figure 26 is an example of a slide that can be inserted after the generic overloading slide (Figure 5) is shown to illustrate a vessel's actual righting arm curves as deck load is increased. The slide refers to a specific vessel for which a new set of stability guidelines had been developed. Initial stability was the same throughout the loading range shown, a crucial point as the crew could not notice any change in the vessel's feel, even though loading varied by more than 16% of the vessel's lightship weight. Clearly, this is a stability weakness for this vessel that can only be determined by calculation, not by at-sea experience. By the inclusion of this slide, the crew understood that the vessel's stability had changed significantly.

Laptop computers are a very common tool of the naval architect these days and work well for making the presentation to individual crews. The same computer, coupled with a video projector, will allow presentations at large gatherings such as trade shows. In the event that a computer is not available on site, individual slides can be printed on letter- or ledger-sized paper ahead of time and displayed on an easel. This paper presentation method can be further enhanced by laminating the pages for added durability.

Adaptability

One of the unique features of the SNAME-developed stability training program is that, with minor modifications, it is suitable for many other types of maritime craft, large and small. Container ships, tankers, passenger vessels, and cargo ships all operate under the same underlying principles of physics. All that is required to adapt this training program is to delete fishing-specific parts, such as towing gear, and modify the illustrations to suit the vessel class. The hands-on demonstration models would require no changes, as they are sufficiently generic in their current form. With these changes, the stability training program would make an excellent refresher course for licensed deck officers and an excellent introductory course for naval architects.

Conclusions and recommendations

Improving the safety of fisherman through educating them about their vessel's stability involves three key topics: stability criteria, stability guidance, and stability training. All three are interrelated. The best stability guidance is of no value if crews have not been educated in how to use it safely. If the stability criteria used to create stability guidance information are substandard, then any stability guidance provided will also be inferior.

With the stability training program developed by SNAME's Ad Hoc Panel 12, dangerous misconceptions about vessel stability that have persisted in the commercial fishing industry can be set straight, and the fishing community can receive knowledge critical to understanding vessel stability. This is the first step in improving the safety of fishermen. The next steps that should be undertaken in future projects involve improving (1) the stability criteria used to develop stability guidance and (2) the methods that can be used to present stability guidance to fishing crew. Completion of these projects will enhance the safety of fishermen.

Acknowledgments

The authors wish to thank the following groups for their valued assistance in the development of this stability training program. Without their comments and suggestions, this program would not been able to achieve its goals.

Members of SNAME's Ad Hoc Panel 12, Fishing Vessel Operations and Safety

Members of SNAME's T&R Panel O-44, Marine Safety and Environmental Protection

U. S. Coast Guard, Headquarters, Fishing Vessel Safety Division

U. S. Coast Guard, regional fishing vessel safety coordinators

Members of the Maritime Lawyers Association Fisheries Committee

The crew and owners of the F/V *Christi Caroline*, F/V *Miss Beth*, F/V *Judy Marie*, F/V *Lisa Kim*, and the F/V *Capt Joe*

The various naval architects and shipbuilders both within and outside the fishing community who reviewed our project

References

Francescutto A (2002). Intact stability: The way ahead. Sixth International Ship Stability Workshop. Webb Institute.

Grochowalski S (1989). Investigation into the physics of ship capsizing by combined captive and free-running model tests. Society of Naval and Marine Engineers *Transactions*.

Herbert J (2002). Progress in prevention and response in fishing vessel safety. *In* Proceedings of the International Fishing Industry Safety and Health Conference (Woods Hole, Massachusetts, Oct. 23-25, 2000), Lincoln JM, Hudson DS et al., eds. Cincinnati, OH: National Institute for Occupational Safety and Health. DHHS (NIOSH) Pub. No. 2002-147.

International Maritime Organization (1995). 1993 Torremolinos Protocol and Torremolinos International Convention for the Safety of Fishing Vessels.

Johnson B, Wallace D, and Savage R (2002). Developing the foundation for an interdisciplinary approach to improving fishing vessel safety. *In* Proceedings of the International Fishing Industry Safety and Health Conference (Woods Hole, Massachusetts, Oct. 23-25, 2000), Lincoln JM, Hudson DS et al., eds. Cincinnati, OH: National Institute for Occupational Safety and Health. DHHS (NIOSH) Pub. No. 2002-147.

Johnson B and Womack J (2001). On Developing a Rational and User-friendly Approach to Fishing Vessel Stability and Operational Guidance. 5th International Ship Stability Workshop, Trieste, Italy.

Lang JS (2002). Fishing vessel safety: A marine accident investigator's perspective. *In* Proceedings of the International Fishing Industry Safety and Health Conference (Woods Hole, Massachusetts, Oct. 23-25, 2000), Lincoln JM, Hudson DS et al., eds. Cincinnati, OH: National Institute for Occupational Safety and Health. DHHS (NIOSH) Pub. No. 2002-147.

Turner J and Petursdottir G (2002). Safety at sea for fishermen and the role of FAO. *In* Proceedings of the International Fishing Industry Safety and Health Conference (Woods Hole, Massachusetts, Oct. 23-25, 2000), Lincoln JM, Hudson DS et al., eds. Cincinnati, OH: National Institute for Occupational Safety and Health. DHHS (NIOSH) Pub. No. 2002-147.

Umeda N, Matsuda A et al. (1999). Stability assessment for intact ships in the light of model experiments. *Journal of Marine Science and Technology* 4: 45-57.

US Coast Guard, Fishing Vessel Casualty Task Force (1999). *Dying to Fish, Living to Fish*. Washington, DC.

Wagner B (2002). Safety and Health in the Fishing Industry: An ILO Perspective. *In* Proceedings of the International Fishing Industry Safety and Health Conference (Woods Hole, Massachusetts, Oct. 23-25, 2000), Lincoln JM, Hudson DS et al., eds. Cincinnati, OH: National Institute for Occupational Safety and Health. DHHS (NIOSH) Pub. No. 2002-147.

Womack J (2002). Small commercial fishing vessel stability analysis: Where are we now? – Where are we going? Sixth International Ship Stability Workshop. Webb Institute.

Appendix: Evolution of stability training

Early stability training
Since the beginning of mankind's venturing to sea, the traditional method of training mariners in ship stability was purely by hands-on experience under the tutelage of experienced captains. Captains were experienced simply because they had survived their previous trips. It was during these long apprenticeships from deckhand to captain that the mariner to-be would learn the proper feel of the sailing ships of the day and thus would be able to command ships.

Although crude, no other practical means of stability training was available. Knowledge of the mechanics of ship stability was in its infancy, and no practical means for the calculation of stability was known. However, given the

Two factors were a key to this success: How ships were propelled and how they were designed. First, sailing ships were always undergoing a simplified inclining, or stability, test while they were underway. Based on the vessel's heel, i.e., freeboard, and current and future weather conditions, sail would be added or removed to maintain an adequate level of safety. Interestingly, the level of safety could be varied based on the weather conditions. In good weather, with steady winds and moderate seas, a lower inclined freeboard would be acceptable and thus higher speeds could be maintained. And when bad weather was encountered, with strong and/or gusty winds and heavy seas, a higher-heeled freeboard would be maintained to enhance safety levels.

Secondly, the design of commercial sailing ships allowed feel to be used to gauge stability. Hull and sail configurations were similar among vessel classes so that the feel of one vessel could be used on another with minimal danger. The evolution of sailing ship design was slow, and any changes typically had small impacts on a vessel's stability characteristics.

Thirdly, with high freeboards and small deck hatches on centerline, the typical commercial sailing ship had relatively high angles of down-flooding. This coupled with a sail's natural tendency for self-dumping the capsizing force as the vessel heeled over, capsizing and sinking from unexpected wind gusts or large rogue waves were rare.

Lastly, only two principal capsizing forces—wind and waves—acted on the typical commercial sailing ship, both of which were reasonably predictable in their effects. Sailing ships did not lift weights or tow heavy gear while under way. They also did not have significant changes in displacement or changes in the center of gravity during their voyages. Free-surface effects were a minimal problem as the sailing ships had inconsequential tankage, nor did they have large deckhouses that might trap large amounts of water high on the vessel. Thus the stability characteristics of a sailing vessel were fairly constant throughout a voyage.

The one class of sailing ship that did not fall into this generalization was large warships. Warships had numerous nontight gun ports near the waterline, which compromised the watertight envelope; they had large weights located high on the hull; and they often pushed the design envelope to meet the grandeur required by ruling monarchs. A few of them were also prone to rolling over in the harbor as well. Interestingly, many of today's commercial fishing vessels have modern versions of these faults with some new ones thrown in.

Why past stability training does not work for today's commercial fishing vessels

Using the stability training methods that were once adequate for sailing vessels for today's fishing vessel is a dangerous proposition. The factors that allowed stability training on sailing vessels through hands-on experience are no longer present on today's commercial fishing vessels.

First and foremost, the use of internal power sources for propulsion removed the continuous inclining test that sailing ship masters could use to gauge their vessel's stability and subsequently make suitable adjustments to its heeling characteristics. Changes in propulsion levels generally do not create noticeable heel changes or changes in a fishing vessel's feel. Neither are propulsion forces self-dumping until severe and often fatal heel angles have been reached.

Second, fishing vessels have been designed with many different configurations to reflect various fishing techniques and local conditions. Safe experience in one fishery cannot be reliably transferred to another fishery having different vessel types and gear. Even within a fishery this problem may exist as some vessels are built to a specific purpose, while others have been converted from other services. An example is the US mid-Atlantic surfclam and ocean quahog fishery, which uses purpose-built vessels, converted Gulf Coast shrimpers, and converted offshore supply vessels. This fishery also uses several fishing methods, side-rigged and stern-rigged dredges, and may carry the heavy catch in fish holds, totally on deck, or a combination of the two.

Third, the general trend in fishing vessel design has been toward making vessels more sensitive to stability problems. Freeboards have been shrinking, making shipping water on deck a more common occurrence. Coupled with numerous openings in the watertight envelope for fish hold hatches and personnel access now closer to the waterline, the danger of down-flooding has increased.

Fourth, numerous and highly variable capsizing forces now act on commercial fishing vessels that were not a concern to sailing masters. Fishing vessels may tow large nets, handle heavy crab pots over the side, or lift heavy weights while fishing. And these capsizing forces do not naturally diminish as the vessel heels over. In fact, it is not unknown for purse seiners to capsize when the fish in the net are spooked and dive en mass, literally pulling the vessel over.

Lastly and most importantly, the stability characteristics of most commercial fishing vessels change significantly throughout a voyage. Fuel and other consumables are burned off, and the catch is brought on board, resulting in significant changes in displacement and the vessel's center of gravity. Since both of these factors are the dominant influences on a commercial fishing vessel's stability, stability characteristics will vary greatly during the voyage. This creates a constantly moving target with regard to the vessel's feel that the crew can not reliably predict. In simple terms, a good feel now may not be correct at a later stage in the fishing trip.

Stability myths and misconceptions

The design and operational characteristics of today's fishing vessel has led to the creation and subsequent handing down of many myths and misconceptions about how stability works and how the feel of a vessel can be used to judge if its stability levels are adequate. Typical misconceptions include that the feel of the vessel will allow a captain and crew to gauge the vessel's stability levels or that the use of paravanes (flopper stoppers) will improve a fishing vessel's stability and its rolling behavior or that adding ballast low will automatically improve stability. Interestingly, many of these misconceptions date back from the sailing days and have been handed down from captain to captain over the decades, if not centuries.

These misconceptions continue to appear in trade journals, books about fishing, and in general talk around the docks. This is not the fault of the fishermen, but the continuing failure of naval architects, safety trainers, and regulators to address stability training adequately to crews. Existing stability training programs generally contain one or more fundamental flaws that limit their use in teaching the basic concepts of stability that would put these misconceptions to rest.

Modern commercial fishing vessels have very complex stability characteristics. With the many different types of fishing vessels, fishing methods, and loading conditions, any stability training program must be carefully developed to cover all potential scenarios crews might encounter. No longer can stability be taught by the traditional method of hands-on experience handed down from captain to crew.

Vessel Safety
Stability training program for commercial fishing vessel crews

SESSION EIGHT:
WORKERS IN FISH PROCESSORS

Gloves and boots drying in Norway (Photo courtesy of R. Evans and B. Bang)

Session Eight

WORK ENVIRONMENT AND HEALTH IN THE SEAFOOD-PROCESSING INDUSTRY IN NORTHERN NORWAY

R. Evans, B. Bang,[1] L. Aasmoe,[1] R. Boe, G.S. Andorsen, B. Aamodt,
E. Kramvik, L. Aardal, B. Pedersen, T. Rasmussen, H. Rasmussen, C.
Egeness, T. Eriksen, I. Espejord, A. Bjørnbakk, and M. Johnsen

Department of Occupational and Environmental Medicine
University Hospital, Northern Norway
Tromsoe, Norway

[1]Project leaders

Introduction

The seafood processing industry is the second largest export industry in Norway. Thus, the industry plays a very important role in the cultural and economic life of coastal Norway. Percentage of days lost to sick leave in the fisheries industry for office personnel was 4% in 1998 and 1999, as opposed to 12% and 11% for production workers for the same years (Bang et al. 2000). Yet little information has been published on the relationship between working environment and worker health. Information on preventive measures and the feasibility of such measures is also lacking

Since 1999, senior researchers of the Department for Occupational and Environmental Medicine at the University Hospital of Northern Norway have led a series of projects on seafood processing plants in northern Norway (Bang et al. 2000; Bang and Aasmoe 2002). These projects are being funded in part by the Confederation of Norwegian Business and Industry. The goals were to obtain hard data and specific knowledge of the work environment and worker health in the seafood processing industry and return the knowledge acquired back to that industry. The strategy chosen was to do a series of studies, refine the areas of interest and methods used, and present the results to industry.

This paper focuses on the development of input to the industry. Articles presenting hard data and results are in preparation and will be published elsewhere (Bang et al. ; Aasmoe et al.).

Methods

The series of studies included a literature review, pilot study, main study, and presentation of findings to industry. The literature review pointed out both probable environmental factors and health effects of interest. Environmental factors include noise, thermal environment (Kuklane 1999), finger temperature (Halkier-Sørensen and Thestrup-Pedersen 1991), air quality (Orford and Wilson 1985; Malo et al. 1988; Douglas et al. 1995; Jeebhay et al. 2000), contact with seafood proteins (Schwartz and Tabershaw 1945; Kavli et al. 1985; Bjelland et al. 1989), repetitive movements, ergonomics, and work organization (Pålsson et al. 1998). Possible health effects were skin irritation, respiratory symptoms, allergy symptoms, hearing loss, and muscle and joint pain.

The pilot study was carried out in five whitefish processing plants. It included 356 returned questionnaires, including ones translated into Tamil, and a health study of volunteers on lung function, allergies, and general antibody level (IgE). The findings confirmed that cold, noise, heavy and/or repetitive work, organizational factors, irritative/allergenic exposure, and air quality are important factors in worker environment and health.

For the main study, a new questionnaire was designed and applied to whitefish, salmon, herring, and shrimp processing plants. Subgroups were formed to look at the following: skin and respiratory symptoms, musculoskeletal symptoms, thermal environment, and noise. As the project progressed, a subgroup was found to be necessary to look at the use of diesel- and propane-powered forklifts indoors and effects on air quality. Questionnaire results were used by all subgroups.

On-site data collection was carried out in two ways: environmental mapping at 17 (in some cases 19) plants and health studies of more than 220 volunteers. Environmental mapping included monitoring bioaerosols, mold, total microorganisms, endotoxins, exhaust gasses, noise levels, thermal environment (including information on the construction and heating of the plants, relative humidity, and ambient and worker temperatures), and 16 in-depth interviews on ergonomic and organizational aspects.

Health studies included spirometry and peak flow measurement series over 2 weeks, as well as the collection of blood samples for allergy tests.

Results: Return to the industry

The product returned to the industry was a folder and educational materials. The folder presents the findings in a constructive manner, i.e., as practical, and for the most part, low-cost measures that will better the working environment. The folder should be in use by early 2004.

Educational materials are being developed for use by industrial hygienists. The results for each of the main topics of interest—thermal environment, noise level, ergonomics, work organization, air quality—have been summarized with references. The plan is to test these materials in the field early in 2004. Information from all subgroups is included. Two of the main points we hope to share are that—

1. The total working environment and the interaction of factors must be considered. Suggestions for improvement must be economically feasible and be part of developing job security for the workers.
2. It is not unethical to point out that a good working environment quite often means a less expensive working environment in the long run. As an example, one may look at suggestions for improving air quality. These include—

    + Replace propane- or diesel-powered forklifts with electric forklifts.
    + Do not use ventilation systems that recycle air.
    + Enclose production areas where bioaerosols are formed, such as where high water pressure is used to peel shrimp.
    + Reduce the of likelihood of mold and bacteria growth by reducing water and waste spill and providing good ventilation.

Improved air quality also means that a better product and better worker health may ensue. Thermal environment is closely related to air quality and product quality. The thermal environment must be such that the seafood being processed is kept refrigerated, but the workers must not be so cold that they cannot do their work. Ambient temperature may also contribute to airway problems. Temperature will be influenced by the placement of heat sources, ventilation currents, loss of heat to evaporation of waste water, loss of heat to refrigerated areas and outdoors, and so forth. All of these factors provide opportunities for improvement in most plants. Again, less spillage means less heat lost to evaporation, less risk of falls, better air quality, and possibly lower worker insurance rates and lower electricity bills.

How workers experience the working environment is definitely related to organizational factors. Cold workers hurt more. Job rotation can mean moving around and staying warm. Perception of the work environment is important. Knowing how best to do your job lessens muscle tension. Mentoring new employees by older, good workers seems to be a definite asset for all. The pleasure of good colleagues and a management that is responsive to workers aids a positive perception of the work environment.

References

Aasmoe L et al. (In prep). Skin symptoms in the seafood-processing industry in northern Norway

Bang BAK, Bjørnebakke R et al. (2000). Arbeidsmiljø og helse i 5 fiskeribedrifter i Nord Norge (In Norwegian,Work environment and health for 5 fish processing plants in northern Norway). Report Regional Hospital, Tromsø, 207 pp.

Bang B and Aasmoe L (2002). Arbeidsmiljø og helse i fiskeindustrien i Nord-Norge. (In Norwegian, Work environment and health in the seafood processing industry in northern Norway). Report, University Hospital of Northern Norway.

Bang et al. (In prep). Exposure and airway effects as recorded for seafood industry workers in northern Norway.

Bang et al. (In prep). Cold work in the seafood-processing industry in northern Norway

Bjelland S, Hjemeland K, Volden G (1989). Degradation of human epidermal keratin by cod trypsin and extracts of fish intestines. *Archives of Dermatological Research* 469:73.

Douglas JD, McSharry C, Blaikie L, Morrow T, Miles S, Franklin D (1995). Occupational asthma caused by automated salmon processing. *The Lancet* 346(16):737-40.

Halkier-Sorensen L and Thestrup-Pedersen K (1991). The relationship between skin surface temperature, transepidermal water loss, and electrical capacitance among workers in the fish processing industry: Comparison with other occupations. A field study. *Contact Dermatitis* 24:345.

Jeebhay MF, Lopata AL, and Robins TG (2000). Seafood processing in South Africa: A study on working practices, occupational health services and allergic health problems in the industry. *Occupational Medicine* 50(6): 406-413.

Kavli G, Gram IT, Moseng D, and Ørpen G (1985). Occupational dermititis in shrimp peelers. *Contact Dermatitis* 13: 69-71.

Kuklane K (1999). Footwear for cold environments: thermal properties, performance and testing. *Arbete och Hälsa* 23, 89 pp.

Malo JL, Cartier A, Ghexo H, Lafrance M, McCants M, and Lehrer SB (1988). Patterns of improvement in spirometry, bronchial hyperresponsiveness, and specific IgE antibody levels after cessation of exposure in occupational asthma caused by snow-crab processing. *American Review of Respiratory Disease* 138:807-12.

Orford RR and Wilson JT (1985). Epidemiologic and immunologic studies in processors of the king crab. *American Journal of Industrial Medicine* 7: 155-169.

Pålsson B, Strömberg U, Ohlasson K, and Skerfving S (1998). Absence attributed to incapacity and occupational disease/accidents among female and male workers in the fish-processing industry. *Occupational Medicine* 48(5): 289-295.

Schwartz L and Tabershaw IR (1945). Dermatitis in the fish industry. *Journal of Industrial Hygiene Tox* 27(1):27-30.

SESSION NINE:
NEW APPROACHES TO SAFETY TRAINING

Fishermen in Sri Lanka (Photo courtesy of G. Piyasena)

STABILITY TRAINING FOR FISHERMEN: ANCHORED INSTRUCTION USING NARRATIVE THEORY AND VIDEO FOR LEARNING TRANSFER

Capt. Barbara L. Howe, MEd , ONI (Canadian)
Quinte Marine Services Ltd.
New Westminster, British Columbia, Canada

Introduction

A fishing vessel capsizes. As the memorial bells toll for those who perished, the litany for more safety training resonates to the tragedy. The fishing community mourns, fishermen talk amongst themselves about what happened, and the agencies responsible for safety training for the fleet again address the conundrum of why fishermen don't seem to be getting the safety message about vessel stability.

The Canadian fishing vessel *Pacific Bandit* capsized in 1995. One fisherman perished. The Transportation Safety Board of Canada (TSB) recommended that the "Department of Transport, in conjunction with other government departments, agencies, and organizations immediately undertake a safety promotion program for operators and crews of small fishing vessels to increase their awareness of the effects of unsafe operating practices on vessel stability" (TSB M95W0005).

In 1997, the *Pacific Charmer* capsized during a herring opening. Two fishermen perished. The TSB Marine Occurrence Report resulted in changes to the Department of Fisheries (DOF) entry conditions for future vessel selection lotteries for targeted fisheries. In particular, the conditions for entry incorporated requirements regarding the certification of vessel operators (TSB M97W-236).

In a paper presented at SARSCENE about the loss of the *Pacific Charmer*, it was suggested that "educators should focus on the way people construe the world differently, and develop educational concepts and processes tailored to the multiple realities inhabited by fishermen" (Boshier, 1999, p. 70). Boshier located these realities in four conceptual frameworks–functionalist, humanist, radical humanist, and radical structuralist. He argues that using these perspectives to examine fishing vessel incidents is informing with education, rather than "de-contextualized training" and concludes that if continued training efforts are "more of the same, it is a waste of time."

The *Cap Rouge II* capsized in August of 2002. Five persons perished. The TSB report on this incident is in its final stages of completion. It will likely contain recommendations encouraging cooperation among operators of small fishing vessels, government regulators, and marine educators to create an everyday reality where stability is a central and routine consideration in making any decision affecting a fishing vessel's movement and operations.

There would seem to be little disagreement about the need for stability training and education. Whether it be stability awareness efforts with pamphlets, formal certification programs, education nested in complex sociological paradigms, or an every day central focus on stability and fishing vessel operations –the way stability information has been presented to fishermen up until now has not substantially reduced the number of capsized vessels.

A complex problem

Why existing stability training and safety awareness efforts do not seem to be working is a complex problem that, among other factors, involves the interrelationship among fishermen, traditional training methods, and the learning environment. More attention needs to be directed at investigating how people, structure, and culture interact to form a learning context (Merriam and Caffarella 1991).

The fishing industry is typified by rugged individualism. There is an apparent discomfort with the educational setting and a natural reluctance on the part of fishermen to attend formal training (Petursdottir, Hannibalsson and Turner 2001). Fishermen have been described as being outside mainstream culture (National Research Council 1991). Not only are they outside mainstream culture, but the literature suggests that commercial fishing is its own culture, subculture, or community (Anderson and Wadel 1972; Barth 1966; De Santis 1984; Knudsen 1987; Orbach 1977; Pollnac, Poggie, and VanDusen 1995). The individualism inherent to fishermen is not likely to change; it is part of the culture of the commercial fishing industry.

This could in part explain discomfort in traditional educational settings and reluctance to attend training programs. Simply stated, there is a "disconnect" between fishermen's cultural reality and ways of knowing, and mainstream teacher-centered training offered in rectangular classrooms with fluorescent lighting. Adult education needs to be culturally relevant. It has been suggested that "effective learning…demands that adult educators reorient educational practices to incorporate learners' culture into the educational process" (Guy 1999, p. 16). Incorporating cultural aspects in the educational process is to establish an authentic context to the learning experience.

The need to include cultural aspects in the educational process is described by Petursdottir, Hannibalsson, and Turner (2001), who state that "any mandatory [training] programme is likely to be resented, resisted and probably will fail unless it has the support and involvement of fishermen. It is important to offer the training in a realistic environment involving the fishermen in hands-on participation with active feedback" (p. 17). This kind of training is learner focused, exploratory, and interactive with less emphasis on the instructor providing knowledge.

Required training programs traditionally consist of curriculum designed by "experts," presented in lectures, overhead transparencies (now being replaced by PowerPoint), textbooks, workbooks, and videos. What is noticeably missing is the active involvement of fishermen.

"Many training videos are simply lectures transferred to a video format in an attempt to provide information to students" (Learning Technology Center 1992, p. 1). These videos are generally de-contextualized representations of facts and bear little relationship to learners' lives.

Teaching materials and methods that do not relate to learners' life experiences become irrelevant and ineffective in facilitating learning (Guy 1999). Classroom learning is substantially different from a natural learning environment. To contextualize training and education, there needs to be a connection with the fishing culture and the life experience of fishermen. An experience-based pedagogy where fishermen connect new knowledge with lived experience can be facilitated by the use of narrative.

This paper suggests that a narrative perspective can be used in the classroom and with videos to promote involvement and relevance and to create a realistic contextual milieu for learning. Narrative is a way for the fishing culture to be incorporated in the educational process and make subject matter relevant to fishermen. A stability course or training program for fishermen using narrative to shape content, method, and learning transfer will be discussed.

The narrative perspective

"Narrative and stories in education have been the focus of increasing attention in recent years" (Rossiter 2002, p. 1). The adult education research community has shown a growing interest in understanding narrative as a basic way of knowing and teaching. There is a recognition of the importance of narrative as a means of informing educational practice (McEwan and Egan 1995). We can begin to appreciate the importance and role of stories in teaching and learning when we recognize how central narrative is to our lives. This section will discuss how the narrative perspective relates to culture, the individual, and learning.

In the broadest sense, culture is a way of making sense of experience nested in shared history. The language of a culture mediates the defining common beliefs, values, customs, acceptable behaviors, and social organization in a meaningful way. Bruner (1986) suggests that we frame these cultural meanings in story form and narrative.

The agnatic nature of commercial fishing allows stories of practical actions and expected outcomes to be passed down through generations by natural learning that takes place in the context of the activity. Tradition and continuity of practice in the commercial fishing industry is embedded in the narrative process of telling how it was and how it is now. When a fisherman answers a question about why they did something the way they did, often the answer is "because we've always done it that way."

This is not an excuse, but rather an *explanation* that reflects a manner of knowing in a legitimate cultural way. Cultural narratives communicate and conserve shared meanings. To participate in a culture is to know and use a range of accumulated and shared meanings (McKewan and Egan 1995). Those who do not belong to the fishing culture but who attempt to encourage fishermen to act differently are confounded by this embedded cultural way of knowing that seems to defy statistical reason and logic sourced in a techno-rational cultural context.

The narrative and its link to shared meanings in a cultural context defines a broad legitimacy of action. Personal narrative, or the telling of stories, is a way people make sense of and give meaning to their life experience. The way we present our personal life with narrative is a particular interpretation of events and experiences which represents the most coherent and satisfactory account (Rossiter 1999). Bruner suggests that "we represent our lives to ourselves as well as to others in the form of narrative" (1986, p. 40). As we change over time, we construct our narrative to reflect new insights, perspectives, and acquired knowledge.

> Fishermen tell compelling stories about the seas and the near misses and the financial bonanza years. These narratives are rich with individual perceptions of how events came to be. It has been said that "to be a person is to have a story... more than that, it is to *be* a story" (Kenyon and Randall 1997, p. 1).

Narrative accounts, whether from the broad cultural realm or an individual experiential story, can provide a basis for learning. "Given the centrality of narrative in the human experience, we can begin to appreciate the power of stories in teaching and learning" (Rossiter 2002, p. 1). Educators need to listen to fishermen's narratives. These stories can reveal significant gaps in knowledge and understanding of stability principles.

Stability information that informs needs to be presented in a way that allows fishermen to make a connection with their existing personal narrative of meaning from experience. As teachers, it is important to leave interpretive space to the learner and allow them to interact with the subject matter in a way that is compatible with their constructed sense of self and understanding. Rossiter argues that "to tell too much, to provide the answers to all questions spoken and anticipated, is to render the active engagement of the learner unecessary" (2002, p. 2).

Narrative, or story telling, is significant in terms of cultural identity, and can be used to capture teachable moments and allow learners to fill gaps in their personal narrative of knowing with new information and knowledge. Involvement in the learning process and participation in learning activities can be realized by allowing stories to be told. Clark states that "the challenge...is to expand our understanding of narrative and explore exactly how narrative can both facilitate and explain the learning process" (2001, p. 89).

One way that challenge has been put into education practice is with Anchored Instruction, which uses narrative as the locus of its theoretical framework.

Anchored Instruction: Theoretical framework

Theoretical frameworks that inform adult education provide an orientation to help investigate and direct how learning takes place. Narrative theory and its relevance to teaching and learning is pivotal to Anchored Instruction, a theoretical framework developed by the Cognition and Technology Group at Vanderbilt University (CTGV) in the late 1980s.

Their investigation began by asking two groups of students to read text containing technical information. The first group was asked to just read the text, and the second group was asked to read it as though they were planning a trip down the Amazon River. The first group could only recall vague information from what they read. The second group was able to remember specific details from the text passage. The significance of this was the demonstrated effect of contextualized learning and the importance of presenting information in a natural setting rather than as isolated facts.

Anchored Instruction is a paradigm for technology-based learning that uses stories, rather than lectures, to provide a macro-context for teaching and learning. The macro-context narrative, which is the "anchor," is a problem-based story presented in a video that is interesting, believable, and relevant to real-world situations (CTGV 1993). Students are asked to solve problems contained in the context of the video narrative. This is accomplished by having students play authentic roles while investigating a situation—identifying gaps in their knowledge, researching information needed to solve the problem, and developing solutions. The teacher is a facilitator and coach for student problem-solving activities.

This methodology offers an environment where learning is contextualized to enhance transferability of knowledge and encourage ownership of learning. The CTGV developed a series of interactive videos called "The Adventures of Jasper Woodbury." Each video episode is a realistic story that culminates in a challenge to be solved rather than a resolution to the story. Information on how to solve the problems posed in the videos is embedded in the story. Students are able to take the embedded information and facts and transform the "mere facts" into valuable knowledge shaped as conceptual problem-solving tools.

The Jasper videos are designed to bridge the gap between natural learning and classroom learning (CTGV 1993). The narrative is an "anchor" that provides a common frame of reference for the learners and the teacher and thus resembles the contextual element of natural learning. The tasks the learners perform are considered to be authentic because they are contained in and arise naturally from the context of the narrative. The knowledge learned is a tool to accomplish the tasks and allows the learner to see the knowledge as valuable, useful, and transferable to other situations.

In traditional training programs "students rarely see the knowledge they learn in class as a tool to solve real-world problems…and often view it as 'school knowledge' and unrelated to their world" (Learning Technology Center 1992, p. 2). This is in part because the cultural background of teachers and learners is often substantially different, and they do not share a common narrative or context for problem solving. Nor do traditional training videos establish an authentic common context with the viewers. When a video is finished, there is nothing to compel the viewer to transport the information in the video into their real world.

Research findings on the Jasper Woodbury videos indicated that Jasper students "performed as well or better on standardized tests, even though the Jasper classes had spent 3 or 4 weeks less on the regular math curriculum. Jasper students scored much higher on planning and subgoal comprehension problems than their control counterparts" (see peabody.vanderbilt.edu/results).

In summary, the underpinning principles of Anchored Instruction are that

1. An anchor should be used to design learning and teaching activities and the activities should be some sort of case study or problem-solving situation that frames the learning in a realistic and authentic setting.
2. Curriculum materials should allow exploration, questioning and involvement by learners, e.g., narratives contained in interactive videos.
3. As a theoretical framework that informs practice, Anchored Instruction and the use of narrative offers interesting insights and valuable potential for teaching stability to fishermen.

"Da Vinci's Inquest" – A video

I spoke at the beginning of this paper about the capsizing of the *Pacific Charmer*. A short time after that incident the award-winning British Columbia television series "Da Vinci's Inquest" aired an episode loosely based on the *Pacific Charmer*. The video is about the capsizing of the fictional fishing vessel *Provider Quest*. It presents technically correct stability information and also captures fishermen's cultural and personal perceptions associated with stability and commercial fishing. This paper suggests that it is difficult to separate the commercial fishing culture from principles of fishing vessel stability. As the narrative of "Da Vinci's Inquest" shows, they are authentically blended.

Following is a list of stability (bold), cultural (italics), and safety/other issues that are evident in "Da Vinci's Inquest." They are in the approximate order they appear in the video.

> *Centers of experience, grandfather, father, and son*
> EPIRBs and response time by search and rescue
> The role of investigative bodies, the Transportation Safety Board
> *Wife of a casualty looking at her husband in a body bag*
> The role of a union safety representative
> *What kind of a guy was the skipper*
> *Was he competent*
> Skipper's boat burned and that is why he is running a company-owned vessel
> Death by drowning or hypothermia
> *Was alcohol a factor*
> One man missing and presumed drowned
> *If you haven't fished you don't understand what it's all about*
> **All the hatches were open**
> **Seals on the hatches were worn**
> **The lazarette was open**
> **Engine room door was open**
> **Scuppers were blocked by gear**
> **The throttle was open with 30 degrees to port**
> *Fishing means that tragedies occur*
> *Skipper's idea to take on another load*
> Quota system
> **Stability data book for sister ship**

The drum was heavier and higher on deck than the one on the sister ship
Vessel heeled to starboard
The water was so cold and I couldn't breath
Human error
I heard pounding from the overturned hull
It's almost impossible to put a survival suit on in the water
Survival suits—two in fair condition, one hooped
We rolled in seconds
We worked 18-20 hours a day
Doors were open that allowed for down-flooding
We have a healthy respect for the sea
What does the inclining experiment tell us
Center of gravity and how it moves
What do the guidelines in the stability book tell us
The boat always had buoyancy
This time the vessel felt sluggish
The "old man" knew the vessel was sluggish to return to the upright
We always left the doors below open—can't be opening and closing all the time
The "old man" doesn't always tell us everything
The "old man" thought something was wrong after the last refit
We were broke and needed the money
Survival suits were stowed in the crew quarters
Superstition—do something a 1000 times and one time your luck doesn't hold
Grampa wouldn't do anything to put us at harm
Fishing is about families
I never even wanted to be on the boat

After listening to the narrative accounts of the capsizing given in "Da Vinci's Inquest," the inquest jury concluded that no one thing caused the capsizing and that it was an accident. It is apparent why a jury would have difficulty isolating a specific cause from the blended narrative that introduces a number of possible contributing factors.

The narrative "Da Vinci's Inquest" talks about important principles of stability and also raises several culturally embedded aspects of commercial fishing that are not quantifiable. What is not quantifiable relies on the context of the whole narrative for meaning. It is precisely this holistic aspect of "Da

Vinci's Inquest" that makes it a realistic, plausible explanation of the multiple factors that led to the capsizing of the *Provider Quest* in the video. Or, perhaps, to the capsize of the *Pacific Charmer*.

"Da Vinci's Inquest" is a narrative that can educate fishermen about stability. It is captivating, believable, and introduces stability terms and principles in an authentic and realistic context. The cultural issues that the video raises will not be unfamiliar to most fishermen. This fictional narrative provides a contextual framework of operating practice through which fishermen can actively and authentically investigate fishing vessel stability.

Implications for teaching

"Da Vinci's Inquest" is not a CTVG Anchored Instruction video with embedded information designed for problem-solving activities. However, any of the issues identified in the video can provide an "anchor" for stability problem solving activities. Here are some suggestions.

Facilitate the exploration of stability with fishermen by allowing time for them to tell their own personal stories about stability experiences and general knowledge about stability principles. Their stories will reflect how they make sense of and give meaning to their experiences with fishing vessel stability.

Before showing "Da Vinci's Inquest," explain that it is loosely based on the capsizing of the *Pacific Charmer*. If relevant, ask what the fishermen know about the *Pacific Charmer* and discuss what they perceive might have been the possible causes of the incident.

Suggest that as they watch "Da Vinci's Inquest," they listen for things that they can personally identify with. After watching the video, let the fishermen deconstruct the overall narrative in terms of what they believe was the cause(s) of the capsizing. Listen to fishermen and loosely categorize what they say into a few general topics. These topics can reflect shared beliefs about stability and the culture of fishing. For each topic identified, ask problem-solving questions. For example, they may be "what if"-type questions that focus the lens of exploration from different angles and help to reveal what lies below the first layer of perception. Or a question might be "how would you have handled that situation if it happened on the boat you work on." Answers to this kind of questioning can reveal multiple perceptions.

The next session might be on the inclining experiment. Watch that segment of the video again and note how the inclining experiment has contextual significance—why was Da Vinci interested in the inclining experiment? The purpose of the inclining experiment in general terms is to find where the center of gravity of the vessel is in relationship to the keel. Introduce basic terms necessary to explain the purpose and usefulness to a skipper that the information from the inclining experiment provides.

If possible, go to a shipyard and as a group conduct an inclining experiment on a fishing vessel with the help of a naval architect. The relevant real-world information that is useful is where the vessel's lightship KG is, not how it is derived mathematically. Afterward, have stability books available and explain that the page describing the inclining test is the documentation of what the fishermen just participated in. Replay the section in "Da Vinci's Inquest" that talks about the stability book and discuss the concept of sister ships and why in time they may no longer be stability sisters.

As a facilitator, there are any number of ways one can use "Da Vinci's Inquest" to teach fishing vessel stability using Anchored Instruction. One problem-solving activity could be "anchored" in what fishermen understand "human error" to be. Or ask the fishermen what they interpret "if you haven't fished, you don't understand what it's all about" to mean.

Other anchors the video provides are "the vessel felt sluggish," and "we rolled in seconds." Investigate the meaning of GM and how this value relates to stiff and tender ships. What would have caused the vessel to be sluggish, and what could have been done to make it less so? The plugged scuppers and open engine room and lazarette doors provide an anchor for learning about the very real danger of free surface.

Stability incidents are seldom caused by one factor. Generally the causes are multiple and the effect cumulative. After the first viewing of "Da Vinci's Inquest," fishermen were asked what they thought caused the capsizing. At the conclusion of Anchored Instruction stability training using the narrative "Da Vinci's Inquest," watch the entire video again. Once again pose the original question to the fishermen—what do you think caused the capsizing of the *Provider Quest*. This activity can reinforce the cumulative nature of instability occurrences and may bring some interesting changes in responses that reflect transfer of learning about stability principles.

"Da Vinci's Inquest" provides many anchors for problem-based and participatory learning activities about fishing vessel stability and cultural issues that
can have an impact on stability. To use it effectively, educators will need to
set lecture notes aside, pull up a chair, and start listening to the fishermen.

Limitations of Anchored Instruction

One of the limitations of Anchored Instruction as designed by CTGV is
that it requires carefully produced videos that contain embedded facts and
information that can be used in authentic problem solving activities. The
narrative video "Da Vinci's Inquest" was not produced with that purpose.
Nonetheless it provides "anchors" for instruction of vessel stability for fishermen in a theoretical framework for learning that is an alternative to traditional training perspectives.

It is a reality that regardless of what theoretical framework and teaching
perspective guides a teacher's practice, the curriculum is likely to come from
elsewhere, teaching resources are limited, and there is little time for course
development. It is not uncommon to find that instruction is to enable
students to pass examinations such as those required for many certificate
programs. Anchored Instruction that develops real-world problem-solving
abilities may not prepare a fisherman to answer multiple-choice questions
that only measure recognition and recall and not real-life problem-solving
abilities.

Pedagogical content knowledge, that is, a teacher's way of knowing the subject matter and the value they assign to the content, can limit the effective use
of Anchored Instruction. A nautical instructor with a deep-sea background
might emphasize longitudinal stability calculations as they relate to loading
a 150-meter cargo vessel because that is the way stability has been useful to
him. However, longitudinal stability is less critical for a 15-meter fishing vessel. More important to a small fishing vessel is transverse stability, particularly because the vessel loads at sea and not alongside a dock.

A teacher with no knowledge of commercial fishing may find facilitating Anchored Instruction with fishermen difficult and experience problems being
authentically responsive to fishermen's personal narratives.

Discussion

The field of adult education is diverse and guided by ongoing research. Research findings reshape existing theories of learning and practice, and this in turn reshapes classroom experiences.

Research using narrative analysis has been used to examine fishermen's individual perceptions of risk and risk-taking as part of the culture of commercial fishing (Acheson 2002; Murray 2002).

If the narratives from research can inform us about how fishermen perceive their world, perhaps an exploration of their own narrative constructs can help them better understand their personal need for more information and knowledge about fishing vessel stability. Anchored Instruction is an enabling theoretical framework that can allow fishermen to examine their stories and identify information gaps—and then fill the gaps with useful, real-world knowledge.

Further research is necessary to determine whether learning transfer of stability principles within the culture of commercial fishing can be enhanced by Anchored Instruction rather than the traditional teacher-focused programs that are the current practice.

There is no question that engendering a sound understanding of fishing vessel stability presents a major challenge for government regulators, educators, and fishermen.

The importance of that challenge is immediately evident whenever a skipper is faced with an on-the-spot decision whether or not to haul aboard one last net full of fish when he knows that the vessel beneath him is already struggling to remain upright.

Let's hope for the skipper's sake, for the sake of his crew, family, and community that he has really learned stability principles and is able to use his knowledge to ensure that he and his crew will be around to fish tomorrow, and for years to come.

Acknowledgments

This author thanks Victoria Acheson for introducing me to the value of fishermen's narratives and stories for research and how research findings can inform education practices. Also my thanks to Capt. Brian Lewis, TSB, who is so clear about the need for government regulators, educational professionals, and fishermen to work together in order to make fishing vessel stability an everyday reality of practice.

References

Acheson V (2002). Fisher's attributed causes of accidents and implications for prevention education. *In* Proceedings of the International Fishing Industry Safety and Health Conference (Woods Hole, Massachusetts, Oct. 23-25, 2000), Lincoln JM, Hudson DS et al., eds. Cincinnati, OH: National Institute for Occupational Safety and Health. DHHS (NIOSH) Pub. No. 2002-147.

Anderson R and Wadel C (1972). Anthropological essays on modern fishing. Newfoundland Social and Economic Papers No. 5. Memorial University of Newfoundland.

Barth F (1966). Models of social organization. Royal Anthropological Intitute Occasional Paper No. 23. Royal Anthropological Institute of Great Britian and Ireland.

Boshier R (1990). Dying to fish, fishing to die: Society, culture and accident prevention. Presentation at SARSCENE, St. Johns, Newfoundland.

Bruner J (1986). Actual minds, possible worlds. Cambridge: Harvard University Press.

Clark MC (2001). Off the beaten path: Some creative approaches to adult learning. *Adult and Continuing Education* 89.

Cognition and Technology Group (Vanderbilt University, Nashville, TN) (1993). Anchored Instruction and situated cognition revisited. *Educational Technology* March.

De Santis M (1984). Neptune's apprentice. Novato, CA: Presidio Press.

Guy TC (1999). Culture as context for adult education: The need for culturally relevant education. New Directions for Adult and Continuing Education (Summer).

Kenyon GM and Randall WL (1997). Restorying our lives: personal growth through autobiographical reflection. Westport, CT: Praeger.

Knudsen P (1987). You take serious what's said in play: Systematic distortion of communication on a fishing boat. Ann Arbor, MI: University of Michigan.

Learning Technology Center (1992). Nashville, TN: Vanderbilt University.

McEwan H and Egan K (1995). Narrative in teaching, learning and research. New York: Teachers College Press.

Merriam SB and Caffarella RS (1991). Learning in adulthood. San Francisco: Jossey-Bass.

Murray M (2002). The use of narrative theory in understanding and preventing accidents in the fishing industry. *In* Proceedings of the International Fishing Industry Safety and Health Conference (Woods Hole, Massachusetts, Oct. 23-25, 2000), Lincoln JM, Hudson DS et al., eds. Cincinnati, OH: National Institute for Occupational Safety and Health. DHHS (NIOSH) Pub. No. 2002-147.

National Research Council (1991). Fishing vessel Safety: Blueprint for a national program. Washington, DC: National Academy Press.

Orbach MK (1997). Hunters, seamen, and entrepreneurs: The tuna seinermen of San Diego. Berkeley: University of California Press.

Petursdottir G, Hannibalsson O , and Turner J (2001). Safety at sea as an integral part of fisheries management. Rome: Food and Agriculture Organization.

Pollnac RG, Poggie JJ, and Van Dusen (1995). Cultural adaptations to dander and the safety of commercial oceanic fishermen. *Human Organization* 52(2).

Rossiter M (1999). A narrative approach to development: implications for adult education. *Adult Education Quarterly* 99(50).

Rossiter M (2002). Narrative and stories in adult teaching and learning. *ERIC Digest*, 241.

Transportation Safety Board of Canada (1995). Sinking of the fishing vessel *Pacific Bandit*.

Transportation Safety Board of Canada (1997). Sinking *Pacific Charmer*, Pylades Channel.

SESSION TEN:
NEW FINDINGS ON FISHING SAFETY

Pakistani fishermen (*Photo courtesy of Dr. Muhammad Khan*)

Session Ten

QUANTIFICATION OF LOW-BACK AND SHOULDER STRESS IN COMMERCIAL CRAB FISHING OPERATIONS

Donald S. Bloswick, PhD, PE, CPE
University of Utah
Department of Mechanical Engineering
50 S. Central Campus Drive
Salt Lake City, Utah, USA
E-mail: bloswick@mech.utah.edu

Bradley Husberg, MPSH, BSN
Centers for Disease Control and Prevention
National Institute for Occupational Safety and Health
Division of Safety Research, Alaska Field Station
4230 University Drive
Anchorage, Alaska, USA

Eric Blumhagen, PE
Jensen Maritime Consultants
Seattle, Washington, USA

Introduction

Musculoskeletal injuries are prevalent in the commercial fishing industry. Alaska Fisherman's Fund data for 1994-1998 indicate that nearly half of all nonfatal deck injuries in the Alaska fishing industry were strains and sprains. Trapp (1994) notes that spinal sprains and strains were the most frequent injuries in most fisheries and presents Fisherman's Fund data from 1982-1984 indicating that spinal sprains and strains account for 25.3% of the injuries to fishers in the Dungeness crab fishing industry and 18.7% of the injuries to fishers in the king crab fishing industry. Data from the Alaska Trauma Registry (ATR) indicate that approximately 5% of nonfatal injuries involving at least 24 hours of hospitalization from 1991 through 1998 could be attributed to strains and sprains (Thomas, Lincoln et al. 2001).

A possible reason for the difference between the Fisherman's Fund and the ATR data is that sprains and strains, even those requiring significant time off work, often do not result in hospitalization. It is also likely that many strains and sprains that would be reported by workers in a traditional shore-based factory are not reported by commercial fishers because of the tendency to not complain, distance to medical care, and share-based compensation plans.

Myers, Klatt, and Conway (1994) also suggest that strains and sprains, particularly related to the low back, are a leading cause of morbidity among fishers and that the application of ergonomic analysis is an important strategy to address this issue. One recommendation by Working Group IV (Prevention of Non-Fatal Work-Related Injuries) from the Second National Fishing Industry Safety and Health Workshop in 1997 is that ergonomic assessments should be performed to address nonfatal injuries among fishers (Klatt and Conway 2000). Fulmer and Buchholz (2000) emphasize that ergonomic training in the fishing industry should convey an understanding of the relationship between work performed and the risk of musculoskeletal injury. It has also been proposed that if deck crews were to condition themselves physically for their activities, there might be fewer related injuries (Jarris 2000).

Biomechanical analysis

Five major factors associated with the risk of musculoskeletal injury at the back and shoulder are—

> The magnitude and direction of the load or force on the hands.
> The posture assumed during materials-handling activities (including twisting of the torso and horizontal distance of the load out from the torso).
> The frequency of the force application.
> The duration of the activity.
> Static postures.

In this paper, stresses associated with the magnitude of the load and posture of the body will be investigated for several tasks commonly required during on-deck crab-fishing operations.

Two concepts must be understood to appreciate the impact of the work environment on musculoskeletal stress at the back and shoulder. These are *moment* and *compressive force.* A moment (or torque) is defined as the effect of a force acting some distance from the point of rotation. The quantification of moment is force magnitude times distance from point of rotation. Compressive force is the force actually pushing bones together at the joint. In the case of the low back, a major risk factor is the total compressive force pushing the vertebrae together in the lower part of the back. This compressive force results from upper body weight and posture and magnitude and direction of the load and force in the hands. The effect of twisting the torso when exerting force with the hands, while less understood than compressive force, appears to increase the risk of injury to the low back. In the case of shoulders, a major risk factor is the moment or torque about the shoulder joint and the posture of the shoulder.

Whole-body biomechanical modeling is a type of ergonomic analysis that quantifies stress on the back and shoulder when a worker exerts forces on an external object with his or her hands. This paper presents biomechanical analyses of several on-deck crab-fishing operations. Inputs to the model include worker posture, worker anthropometry (body weight and height), and hand-force magnitude and direction. Outputs include low-back compressive force and shoulder moment (torque), which are primary measures of injury potential at these two joints. This quantification allows a comparison of musculoskeletal stress to generally accepted norms and facilitates recommendations for task or procedure redesign to reduce this stress.

Back compressive forces of 770 pounds are generally thought to present risk to some workers, and back compressive forces of 1,430 pounds are generally thought to present significant risk to most workers. No established limits are available for shoulder stress; therefore, shoulder moment is often presented as the approximate percentage of males who can be expected to have the strength required to perform a noted task. When this value approaches 100%, the task can be expected to present shoulder stress to many workers. In addition to the computer-based analysis, user-friendly hand-calculation methods to estimate back compressive force and shoulder moment can be requested from the first author.

Biomechanical modeling of crab-fishing operations

Lifting/lowering
When lifting or lowering a given load, the force in the erector spinae muscle and resulting compressive force can be reduced by keeping the torso as up-

Figure 1: On-deck lifting task bending at waist

right as reasonable and comfortable. A crab-fishing task in which the torso is flexed is shown in Figure 1.

For this study, it was hypothesized that most on-deck lifting loads were 100 pounds or less (for example, two boxes of bait). Remember that a compressive force of 770 pounds is thought to present a risk for some workers and a compressive force of 1,430 pounds is thought to present a significant risk for most workers.

Figures 2 and 3 represent the output of a computer-based biomechanical model for a lift with the torso flexed (stoop) and more upright (squat) with a load of 100 pounds. Figure 4 compares compressive force on the back for a stoop lift and squat lift with hand loads of 25, 50, 100, and 150 pounds. As can be seen, low-back compressive forces are lower when lifting with the torso upright than when lifting with the torso flexed forward.

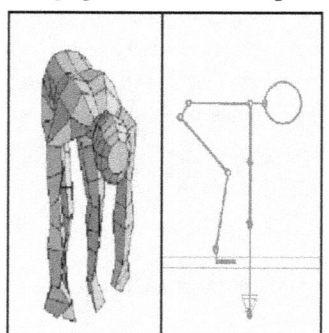

Figure 2: Biomechanical representation of a lifting task with torso flexed (stoop).

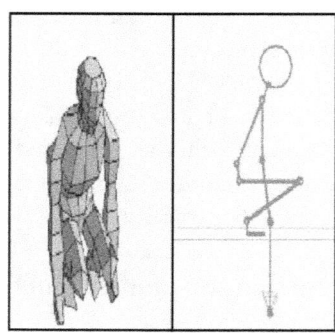

Figure 3: Biomechanical representation of a liftimg task with torso more upright (squat).

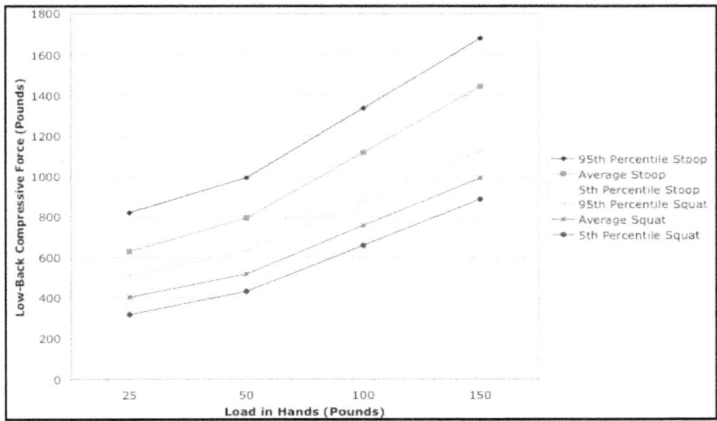

Figure 4: Low-back compressive force for stoop and squat lifting with loads of 25, 50, 200, and 150 pounds

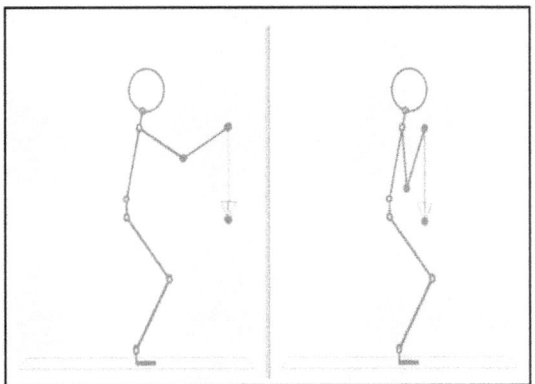

Figure 5: Biomechanical representation of a lifting task with load held out from body and with load held close to body

For a given load, the force in the erector spinae muscle and resulting compressive force can also be reduced by keeping the load close to the body. Figure 5 contains biomechanical representations of a lift task with the load far from and close to the body. Figures 6 and 7 illustrate the compressive force and the percentage of males with the shoulder strength to perform "close lifts" and "far lifts" with load of 25, 50, 100, and 150 pounds. These figures show that the low-back compressive forces are lower and the percentage of male workers who would have the shoulder strength to perform the task is higher when the load is lifted close to the body than when it is lifted held out in front of the body.

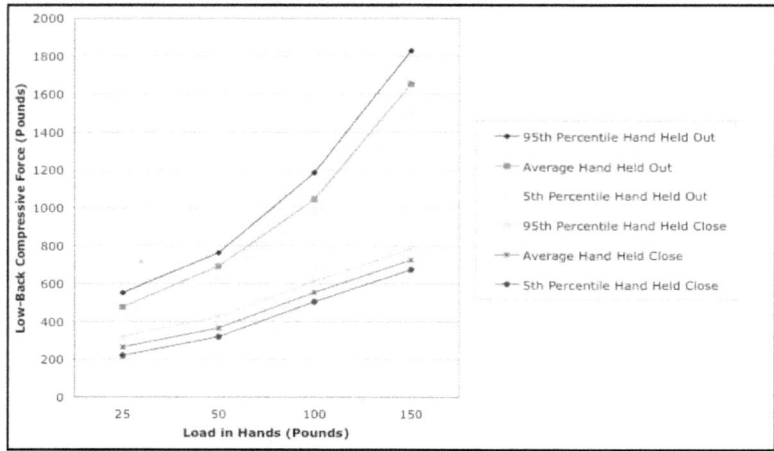

Figure 6: Low-back compressive force for lifting task when load is held out from body and when load is held close to body at loads of 25, 50, 100, and 150 pounds

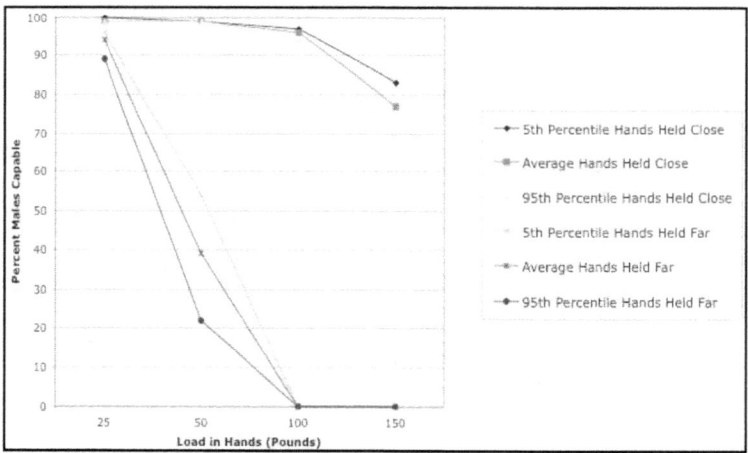

Figure 7: Percentage of male workers with shoulder strength to perform a lifting task when load is held out from body and when load is held close to body with loads of 25, 50, 100, and 150 pounds

For each posture, the three plotted lines represent the compressive force for the 5th, 50th, and 95th percentile male, which approximately represent the smaller/weaker, average, and larger/stronger person. Data for the 95th percentile (larger/stronger male) are included to recognize the fact that commercial fishermen may tend to be stronger and better conditioned than the average industrial population.

Figure 8: Opening crab pot door with hands behind torso

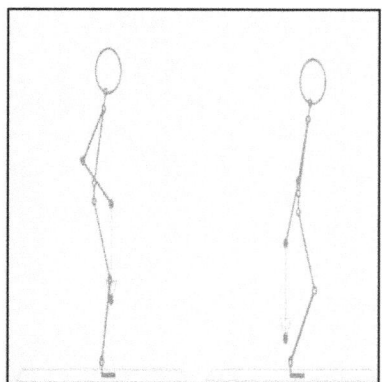

Figure 9: Biomechanical representation of lifting with hands in front of and behind torso

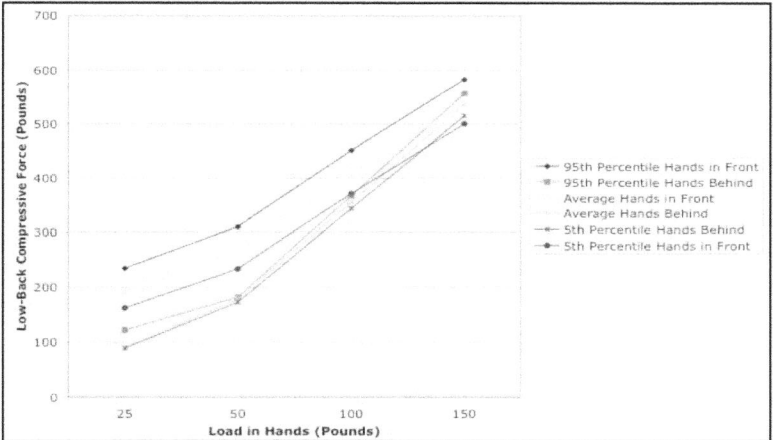

Figure 10: Low-back compressive force when hands are in front of torso and behind torso with loads of 25, 50, 100, and 150 pounds

Low-back compressive force during lifting and lowering can be minimized by locating the hands absolutely as close as possible to the low back. A crab-fishing task in which this is commonly done is shown in Figure 8. Figure 9 contains biomechanical representations of a lift task with the hands in front of and behind the body. Figure 10 compares the compressive force on the back with the hands in front of and behind the body with hand forces of 25, 50, 100, and 150 pounds. This figure shows that the low-back compressive forces are lower when lifting with the hands behind the body than when lifting with the hands in front of the body, even when the hands are right in front of the body.

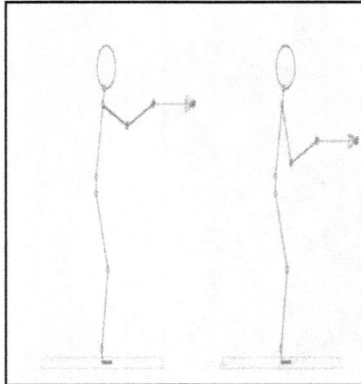

Figure 11: Biomechanical representation of pulling task with hands above waist and close to waist

Push/pull

The generation of pushing and pulling forces also results in external moments that must be resisted by the muscles of the body. For a given push or pull force, the low-back compressive force can be reduced by task design and/or work practices that orient force direction as close as possible to the waist. This reduces the external moment and torso muscle force and, consequently, the low-back compressive force. Figure 11 contains biomechanical representations of a pull task with the force far from and close to the waist. Figure 12 compares compressive force on the back with force far from and close to the waist for pulls with hand forces of 25, 50, 100, and 150 pounds. These forces are representative of the push and pull forces often exerted during crab-fishing operations. For example, pushing or pulling a crab pot weighing 750

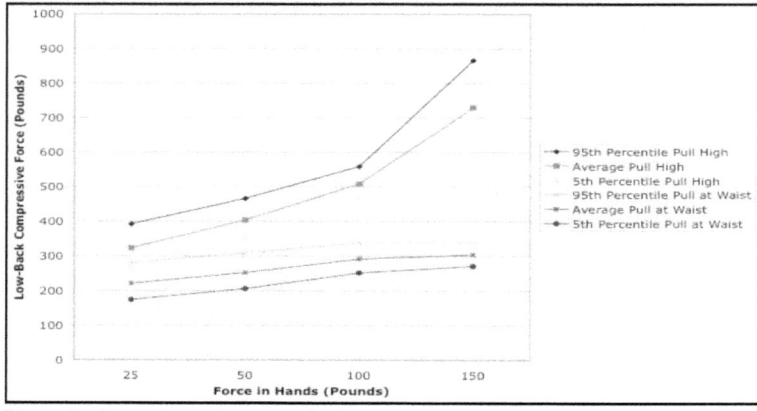

Figure 12: Comparison of compressive force on back with force far from and close to waist

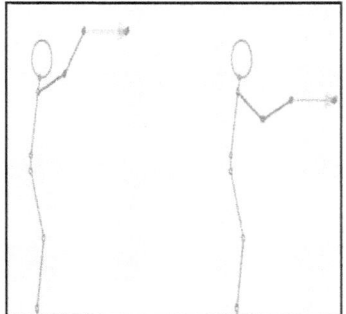

Figure 14: Biomechanical representation of task with hands above shoulder

Figure 13: Pull on rope with hands above shoulder and close to shoulder

pounds on a boat deck with a coefficient of friction of 0.2 would require a pushing force of approximately 160 pounds. The coefficient of friction is the percentage (0.2 = 20%) of load weight that must be exerted horizontally to slide the load. (It should be noted that a worker would have to have his/her feet braced to generate this force without sliding on a deck with a coefficient of friction of 0.2.) It is hypothesized that push/pull forces would seldom, if ever, exceed 175 pounds.

As can be seen in Figure 12, low-back compressive forces are lower when pulling with the line of action of the force close to the waist than when pulling with the lines of action of the force above the waist. The comparative results for a push would be the same.

A major risk factor for shoulder injury is the moment or torque about the shoulder joint that results from the posture and magnitude and direction of the load/force in the hands. For a given push or pull force, the shoulder moment can be reduced by task design and/or work practices that orient the force direction through the shoulder. For example, in Figure 13, a crab-fishing operation is shown where the line of action of the pull force is exerted above the shoulder and where the line of action of the pull force is closer to the shoulder. Figure 14 contains biomechanical representations of these two tasks.

Figure 15 illustrates the percentage of males having the shoulder strength to perform pull tasks with the force close to the shoulder and above the shoulder using hand forces of 25, 50, 75, and 100 pounds. It can be seen that

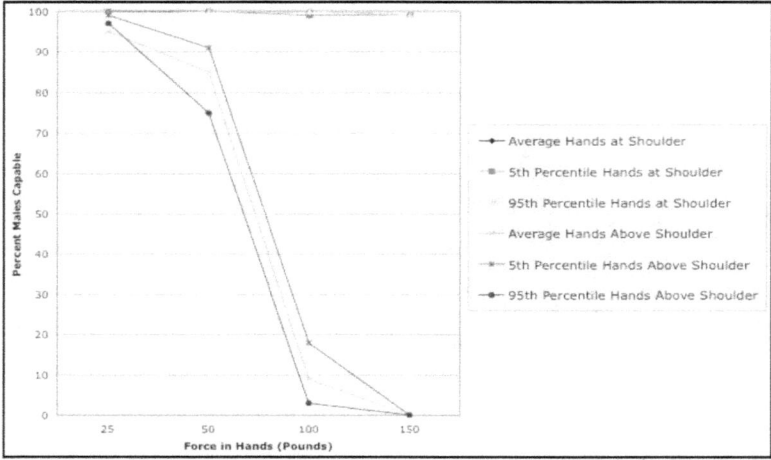

Figure 15: Percentage of male workers with shoulder strength to perform pulling task with hands above shoulders and with hands close to shoulders at forces of 25, 50, 100, and 150 pounds

the percentage of male workers who can be expected to have the strength required to perform the task is considerably higher when the line of action of the pull force is through the shoulder than when it is overhead.

As might be expected, workers often develop work practices that optimize overall stress on the body, but do not minimize stress on any one joint. An example of this is positioning the hands approximately half-way between the low back and the shoulder when pulling or pushing so that neither the back nor the shoulders are highly stressed.

Summary and recommendations

Biomechanical modeling can be used to estimate the musculoskeletal stress in commercial crab-fishing activities. The output of the biomechanical model is an objective method to identify equipment, task, and procedural and design recommendations that can be made to reduce stress to acceptable levels.

Stresses on the low back can be reduced by designing lifting and lowering tasks and work practices so that (1) the torso remains as upright as possible and (2) the hands are held close to the body. For tasks involving the generation of horizontal forces, both pushes and pulls result in less low-back stress when the line of action of the force is approximately at waist level. Lifting "behind the back" results in less low-back stress than lifting in front. Shoulder stress can be reduced by designing pulling tasks and using work practices so that the line of action of the force is approximately at shoulder level. An overall optimization of back and shoulder stress results when the line of action of the push or pull force is approximately at chest level.

In conclusion, while on-deck operations during crab fishing may require considerable hand force to lift, lower, push, or pull objects, musculoskeletal stress and the resulting risk of injury can be minimized if care is given to correct body posture.

Information in this paper can be used to assess the stresses and forces experienced by fishermen performing these or similar tasks in other commercial fishing operations. Future recommendations include collaboration between fishermen and fishing companies to obtain specific measurements of these tasks. The authors are willing to work with any of those parties who might be interested.

References

Center for Ergonomics, University of Michigan (Ann Arbor, Michigan) (2001). 3D Static Strength Prediction Program, Version 4.3.

Fulmer S and Buchholz B (2002). Measuring risk of cumulative musculoskeletal trauma in fishing vessels. *In* Proceedings of the International Fishing Industry Safety and Health Conference (Woods Hole, Massachusetts, Oct. 23-25, 2000), Lincoln JM, Hudson DS et al., eds. Cincinnati, OH: National Institute for Occupational Safety and Health. DHHS (NIOSH) Pub. No. 2002-147.

Jarris R (2000). Optimizing medical care at sea. *In* Proceedings of the Second National Fishing Industry Safety and Health Workshop (Seattle, Washington, Nov. 21-22, 1997), Klatt ML and Conway GA, eds. Cincinnati, OH: National Institute for Occupational Safety and Health. DHHS (NIOSH) Pub. No. 2000-104.

Klatt ML and Conway GA (2000). Working group recommendations. *In* Proceedings of the Second National Fishing Industry Safety and Health Workshop (Seattle, Washington, Nov. 21-22, 1997), Klatt ML and Conway GA, eds. Cincinnati, OH: National Institute for Occupational Safety and Health. DHHS (NIOSH) Pub. No. 2000-104.

Myers ML, Klatt ML, and Conway GA (1994). Executive summary. *In* Proceedings of the National Fishing Industry Safety and Health Workshop (Anchorage, Alaska, Oct. 9-11, 1992), Myers ML and Klatt ML, eds. Cincinnati, OH: National Institute for Occupational Safety and Health. DHHS (NIOSH) Pub. No. 94-109.

Thomas TK, Lincoln JM, et al. (2001). Is it safe on deck? Fatal and nonfatal workplace injuries among Alaskan commercial fishermen. *American Journal of Industrial Medicine* 40(6): 693-702.

Trapp PS (1994). Nonfatal injuries in the Alaska commercial fishing industry. *In* Proceedings of the National Fishing Industry Safety and Health Workshop (Anchorage, Alaska, Oct. 9-11, 1992), Myers ML and Klatt ML, eds. Cincinnati, OH: National Institute for Occupational Safety and Health. DHHS (NIOSH) Pub. No. 94-109.

EFFECTS OF COLD WATER IMMERSION ARE DEADLY BE PREPARED!

Shane K. Neifer
Occupational Safety Officer
Fishing Team Leader
Prevention Division
Workers' Compensation Board of BC
4450 Lakelse Ave.
Terrace, BC, Canada
E-mail: sneifer1@wcb.bc.ca

Introduction

On 9 October 1999, the fishing vessel *Nopsa*, a 16.8-metre (55-foot) black cod longline trap vessel, was near the end of a 3-week trip. It was located on the Bowie Seamount, approximately 90 nautical miles west of the Queen Charlotte Islands on British Columbia's north coast. There was a crew of five on board, and after much bad weather the seas were finally calm. At approximately 2155 hours, the crew had 30 of the string of 60 traps on board. The current trap was somewhat heavier than the previous traps, and two of the deckhands were having trouble stacking it on top of the other traps on deck.

On the second attempt, the trap fell to the deck, causing the two deckhands to scramble to safety. One man moved to the center of the vessel, and the other moved to the starboard rail. The crew member who had moved to the rail of the vessel misjudged his location and slipped forward of the guardrail securing the trap storage area and fell headfirst into the water. He was not wearing personal floatation protection. Although the initial response from the crew was laughter, they quickly realized that the deckhand was in distress and began rescue attempts. The crew member was recovered on board within 9 minutes of entering the water; however, he was unconscious by this time and attempts to revive him were unsuccessful. During rescue attempts, a variety of lines and floating objects were thrown well within reach of the man overboard, but he made no attempt to grab onto any of them and aid in his own rescue. He was able only to continue to cry out for help. The water temperature was between 10° and 11 °Celsius (50° and 52 °Fahrenheit).

On 08 November 2001, the fishing vessel *Golden Dragon*, a 10.7-metre (35-foot) single-pot crab-fishing vessel was in Hecate Straits approximately 34 nautical miles southwest of Prince Rupert on British Columbia's north coast. At approximately 2300 hours, the crew of three were pulling pots at the south end of their line of gear. The winds were from the south and south-west at 12 to 20 knots, making for less-than-ideal but not impossible fishing conditions. The vessel began to roll and pitch, requiring the crew to ensure that everything on deck was well secured. The crew was in the process of emptying crabs from, and re-baiting, the pot on deck when the vessel took a port turn. The crew had been pulling traps on the starboard side leaving the buoy line in the water during the process. The buoy line became caught in the propeller during the port turn, and as the line came tight, it started to pull the partially processed trap off the setting table. One of the crewmen reached to secure the trap as it was prematurely sliding over the side of the boat and was pulled into the water with the trap. As he entered the water, he released the trap and remained at the surface. As in the previously discussed incident, various lines and floating objects were thrown well within reach of the man in the water, but he made no attempts to aid in his own rescue. The stricken crewman was finally recovered on board after approximately 12 minutes in the water. The crewman was unconscious and attempts to revive him were unsuccessful. None of the crew were wearing floatation devices. The water temperature was 9 °Celsius (48 °Fahrenheit).

Both of the above-mentioned incidents were recorded by on-board video equipment used for fisheries management. The video evidence provides a running clock and GPS location data as well as vessel speed and direction.

On 20 September 2000, the fishing vessel *Ocean Jade*, a 15.2-metre (50-foot) salmon troller was at anchor for the evening in Pocket Inlet on the west side of the Queen Charlotte Islands on British Columbia's north coast. A crewman fell overboard and was unable to pull himself back on board the vessel. Attempts to regain entry, with the vessel master's assistance, were un-successful, and the crewman began to panic. The master of the vessel, sens-ing that his crewman was in very real danger of drowning, also entered the water. Neither person was then able to gain re-entry onto the vessel. The master attempted to swim to his brother's vessel that was anchored some 200 metres (650 feet) away. The master did not arrive at his brother's vessel, and his body has never been recovered. The crewman who had initially fallen into the water was recovered almost 10 hours later beside the boat he had fallen from. Although he was severely hypothermic, he survived his ordeal. The water temperature was 12 °Celsius (54 °Fahrenheit). Neither crewman was wearing personal floatation protection.

In another incident, which occurred on 27 March 2002, two fishery technicians were conducting eulachon research in the Skeena River on British Columbia's north coast. Although the portion of the Skeena River on which the technicians were working is not considered "tidal," it is influenced by the tides at the mouth of the river. The water rises and falls on-cycle as water levels change further down the river. The technicians had walked out to an island in the river at low water to conduct part of their survey. When they attempted to return, they found that the river was now too deep to cross back to the mainland. In their initial attempt to recross the river, both men filled their chest waders. The technicians returned to the island cold and wet to wait the 5 hours before the river would be low enough to cross. They had no survival gear and were not equipped to start a fire. After some discussion and against the advice of his partner, one of the technicians decided to swim across the channel. The narrowest part of the channel at that time was approximately 90 metres (300 feet) across. Their truck was parked on the far side of the channel next to the highway running alongside the river. Being in good physical condition and a good swimmer, the technician felt confident he could cross the river to the vehicle and go for help. He removed his chest waders, draped them across his shoulder and began to swim. His partner watched him make excellent progress and decided to also attempt the swim. He later reported that the water was so cold that it hurt too much to continue the crossing, so he returned to the island to monitor the progress of his partner. The swimmer reached within 3 metres (10 feet) of the far shore, at which time he simply sank and subsequently drowned in about 1 meter (3 feet) of water. The river water temperature ranged between 0° and 1 °Celsius (32° and 34 °Fahrenheit). The pathologist reported the incident as a "simple drowning." The author questioned a possible relationship between the drowning and the water temperature. The coroner, however, confirmed the pathologist's diagnosis that there was no evidence of hypothermia, and in fact the event was a "simple drowning."

These four summaries of actual incidents investigated by the author lead him to believe that stating that the casualties simply "drown" was incomplete. He felt that exploring and understanding the physiological aspects of immersion could lead to development of initiatives to reduce or prevent the continued occurrence of this type of accident.

Problem

The significant issue that stood out to the author in the four examples above was that the victims all seemed to "give up" very quickly. In two of the cases, the casualties appeared unwilling to aid in their own rescue, and in the other two, the casualties were observed to be swimming strongly and then just appeared to sink or went missing before arriving at their destinations.

The leading cause of fatalities in British Columbia's commercial fisheries is drowning. Over 85% of all fishing fatalities from the period 1991 to 2001 on Canada's West Coast were the result of drowning. This includes individuals who entered the water, were not recovered, and are presumed drown.

The Workers' Compensation Board (WCB) of British Columbia investigates all occupational fatalities within its jurisdiction that occur in the province to determine causes of the accident. Once the accident cause is known, then the WCB further investigates to determine (1) if there are preventative measures to develop and/or (2) if there were compliance issues that led to the accident. In British Columbia, although the number of people fishing for a living has decreased significantly over the last decade, fishing still remains one of the most lethal occupations under the jurisdiction of the WCB. Figure 1 illustrates participation levels, total injuries claimed, and total fatalities over the 11-year period reviewed.

It is clear from Figure 1 that injury claims in the fishing industry are being reduced, although not as quickly as participation levels are dropping. Unfortunately, the number of fatalities is not following the same marked downward trend. Figure 2 breaks out the significant types of accidents that have resulted in death within the fishing industry. Eighty-six percent of these deaths, including those missing and presumed drowned (50 of the 58 that occurred from 1991 to 2001) were from drowning. Thus, it becomes important to determine what issues underlie this statistic. Only then can work begin to reduce the number of drowning fatalities within the fishing industry.

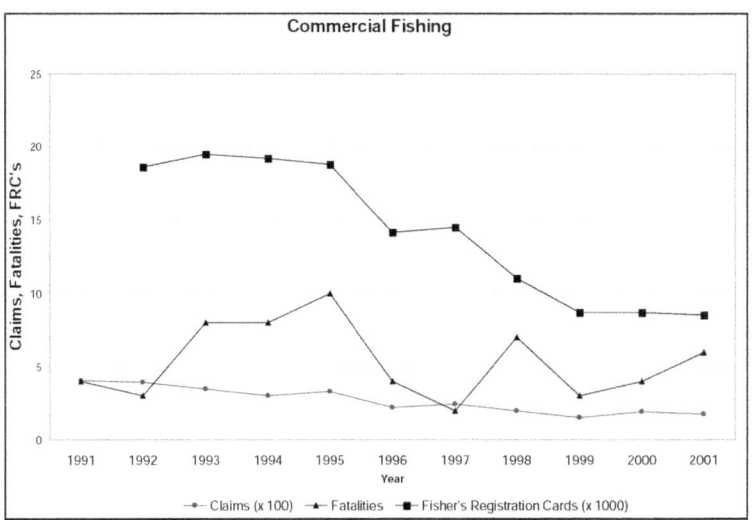

Figure 1: Injury claims, fatalities, and participation in British Columbia's commercial fisheries, 1991 to 2001 (Source: Workers' Compensation Board of BC, Fisheries and Oceans Canada)

Methods

A formal statistical analysis was not within the scope of this report. Rather a review of WCB records involving fishing industry fatalities from drowning and a minimal review of the literature pertinent to cold water immersion were undertaken. Inquires were made to colleagues at the Canadian Coast Guard, Transport Canada, and the Transportation Safety Board concerning their records involving cold water immersion and any literature on the topic. The Transport Canada publication *Survival in Cold Waters* (TP 13833) proved to be an invaluable resource. It can be downloaded from marine publications at the Transport Canada Website (www.tc.gc.ca) (Brooks 2003). Transport Canada commissioned *Survival in Cold Waters* in response to a drowning incident in the Great Lakes in 2000. The report was authored by Dr. Chris J. Brooks of Survival Systems Limited in Dartmouth, Nova Scotia. In this report, Dr. Brooks describes four stages of physiological response by human beings to immersion in cold water.

The author contacted Dr. Brooks to discuss drowning incidents in British Columbia's fishing industry in relation to the stages of cold water immersion. Dr. Brooks then undertook a retrospective analysis of the WCB's investigations into workplace drowning fatalities from 1976 to 2002.

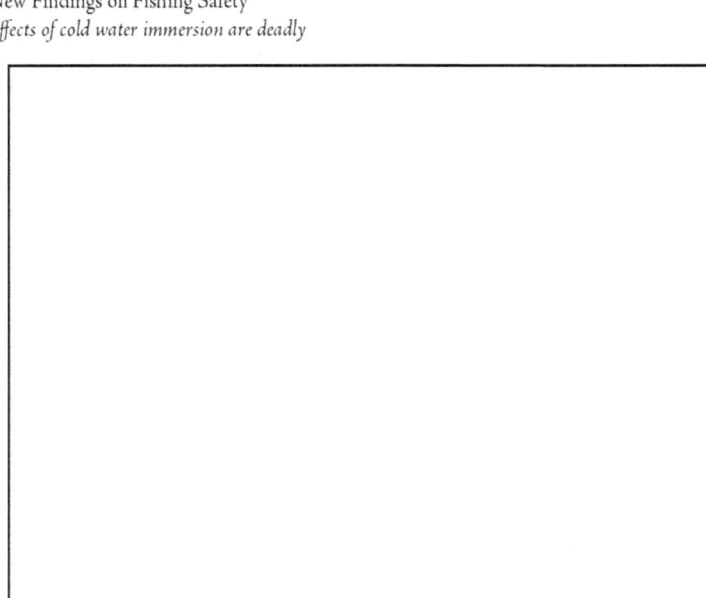

Figure 2: Accident types leading to fatal injuries in British Columbia's commercial fisheries, 1991 to 2001 (Source: Workers' Compensation Board of British Columbia)

Another text was also reviewed. *Essentials of Sea Survival* (2002), authored by Drs. Frank Golden and Michael Tipton, documents historical data through to modern science on the effects of and protection from cold water immersion.

Results

Cold water is defined as water below 25 °Celsius (77 °Farenheit), but significant effects of cold water immersion peak in water below 15 °Celsius (59 °Farenheit). As stated earlier, there are four stages involved in cold water immersion. Stage 1, cold shock or initial immersion, can lead to death within the first 5 minutes. Stage 2, swimming failure or short-term immersion, which can be deadly from 3 to 30 minutes after entering the water. Stage 3, hypothermia or long-term immersion, can take effect and be fatal within 30 minutes. Stage 4, post-rescue collapse, involves fatalities occurring during or after rescue. Outlines of the four stages, as described by Drs. Brooks, Golden, and Tipton, are provided below.

Stage 1- Cold shock
Cold shock occurs immediately upon immersion in cold water. It is a short-lived event, lasting only 3 to 5 minutes. It is the human's physiological

response to sudden cold water immersion. It often leads to panic, which can result in quick drowning. These uncontrollable responses are as follows:

  + A large inspiratory gasp immediately upon immersion.
  + A four-fold increase in respiratory rate.
  + A reduced breath-hold capability.
  + A large increase in heart rate.
  + A large increase in blood pressure.
  + Onset of decreased peripheral dexterity.

The initial inspiratory gasp and increase in ventilation rate can lead to involuntary aspiration of water if the victim's head is below the surface, resulting in quick drowning. The hydrostatic pressure of the water coupled with vasoconstriction of the periphery and increased heart rate quickly increase blood pressure and can lead to heart failure. Heart failure can be fatal on its own, but often just aids in the drowning process. A person's ability to breath hold in cold water is reduced by up to 85%. In air, the average breath-hold capability is just over 1 minute. In cold water, breath-hold capability can be as little as 10 seconds (Golden 2002, p. 63). The effects of stage 1 can easily lead to panic as the person struggles to breathe. In many cases, the result is that the person in the water is unable to function well enough to aid in their own rescue. They may well be unable to grab a line or life ring that has been deployed directly to them as they panic and because the cold water has cooled peripheral muscles, causing loss of dexterity. The human body will adjust to the effects of cold shock quickly. The key is to not inhale water into the lungs and spiral into complete panic within these first 5 minutes. It takes only 150 mL of water in the lungs to drown. Staying afloat by using a personal flotation device, immersion suit, or lifejacket would significantly increase the chance of survival during stage 1 cold shock.

Stage 2 – Swimming failure

Swimming failure lasts from 3 to 30 minutes in cold water. In this stage, the casualty has survived initial immersion and has decided to attempt swimming to rescue. The physiological responses are as follows:

  + Breath-hold capacity remains decreased and continues to decrease as the body remains in the water.
  + Cooling of the external tissues reduces coordination and functioning of the extremities.
  + Breath rate to swimming stroke rate is no longer equal.

+ Swimming angle increases, stroke rate increases, and stroke length decreases.
+ Swimming motion begins to fail as coordination is significantly reduced; signals from the brain do not reach the limbs.
+ Cardiovascular and/or respiratory events can lead to drowning.

One's swimming ability in warm water has little relationship to swimming ability in cold water. In warm water, a swimmer will take one breath per stroke. In cold water, respiratory rate and stroke rate both increase, but not together and quickly diverge. This can enhance the sense of panic. Cooling of the extremities and subsequent decrease in functioning with the increase in stroke rate leads to an increase in the angle of the swimmer in the water. Increasing the angle of the swimmer in the water increases drag and substantially increases the energy required to move through the water. This is happening in a body using all available energy to maintain core body temperature and can lead to a further increase in the feeling of panic, as well as an increase the risk of cardiovascular or respiratory failure. These continuing physiological responses and increasing panic dramatically increase the risk of drowning.

Stage 3 – Hypothermia

Hypothermia is a well-documented physiological event that will be understood at some level by most people in the fishing industry. In cold water, a thermally unprotected individual can die from the effects of hypothermia in approximately 30 minutes. The physiological responses to hypothermia are as follows:

+ Continued shunting of blood away from the periphery to the core.
+ Shivering begins in an attempt to generate heat.
+ Loss or semi-loss of consciousness, leading to drowning.
+ Heart failure.

Hypothermia is generally diagnosed when core body temperature drops at least 2 °Celsius (4 °Fahrenheit). It is usually fatal when core temperature drops by 12 °Celsius (22 °Fahrenheit). The body loses heat four times faster in water than in air. While continued immersion in any water below 35 °Celsius (95 °Fahrenheit) will eventually lead to hypothermia, it is water temperatures below 15 °Celsius (59 °Fahrenheit) that pose the most immediate risks. After surviving the first two stages of cold water immersion, the early onset of hypothermia can lead to a decrease in the will to survive.

There will be a progressive loss of function until the individual loses consciousness and ultimately drowns. Heart failure becomes an issue early in the cooling process as blood vessels constrict, making the heart work harder.

Stage 4 – Post-rescue collapse

Post-rescue collapse is thought to account for as much as 20% of all immersion-related deaths. Issues at this stage occur at the time of or following an individual's removal from the water. The physiological responses to removal from water and rewarming are as follows:

- Significant decrease in blood pressure in the core.
- Aspirated water can degrade lung tissue, affecting the oxygen transfer process.
- Internal injuries sustained during the accident can impact survival.

There are three conditions when this stage most often occurs. The first is when the core blood pressure will decrease suddenly and significantly in response to the vasodilatation of the peripheral blood vessels with removal from the water's hydrostatic pressure and rewarming. This decrease in blood pressure can lead to sudden failure of the heart or brain due to oxygen deprivation. Rewarming should be directed at the core, ensuring that the person is treated much as they would be for shock. This would include maintaining a horizontal position to avoid blood pooling in the lower extremities.

The second condition of concern is when water has been aspirated into the lung tissue, initiating several physiological events. The alveoli are degraded, and oxygen exchange can be seriously compromised. This can lead to death many hours after rescue as inflammation and continued tissue damage result in a pneumonia-type response. This must be treated medically.

Lastly, internal injuries and head and neck injuries can easily occur as someone fell overboard. Internal injuries can lead to fatal bleeding as rewarming stimulates the flow of blood throughout the body.

Summary

Knowing and understanding the four stages of cold water immersion and their effects can lead to the development of protocols to protect individuals in the fishing industry. It is very likely that the casualties from the *Nopsa* and the *Golden Dragon* succumbed to drowning as a result of cold shock (stage 1). It also appears that the master from the *Ocean Jade* and the fishery technician on the Skeena River were classic examples of swimming failure drowning in stage 2 of cold water immersion.

Although hypothermia is commonly recognized as a potentially fatal event and does account for many deaths, in actual fact it may not be responsible for as many deaths as cold shock and swimming failure. Surviving the first two stages will lead to hypothermia becoming a serious risk where drowning may result even with the use of floatation devices. The exception is the use of a survival or immersion suit, which will add many hours to survival time in cold water.

In recent decades, many fishermen have become well-informed on the physiology of hypothermia, and they will contend that immersion in cold water for a short period of time is not terribly risky. There are countless anecdotal accounts of people falling overboard and being pulled safely back onto the vessel within a seconds or short minutes. What is not understood is that a seemingly insignificant variation in an identical situation could have fatal results. For example, an occasional occurrence among the herring and salmon seine fisheries is having a crewman go overboard clinging to the net twine. If the individual's head never submerges or he never lets go of the net, he may be retrieved with little trouble. However, all it takes is for the person's mouth to be below the water when cold shock causes the initial inspiratory gasp reflex and the outcome can be substantially more catastrophic.

Even the abbreviated literature review conducted for this report substantiates the events leading to drowning during cold water immersion when related to actual fatal incidents in British Columbia's commercial fisheries. Understanding the very high risk potential that cold water immersion poses and giving it the concern it deserves will motivate masters and crew to arrange work spaces and wear appropriate personal floatation protection to minimize the risk of these types of injuries. Masters and crew will also develop and practice procedures to quickly recover a crew member overboard on the first attempt!

Conclusions

Fishermen on Canada's west coast have long known the effects of hypothermia. This is a result of tremendous educational initiatives from a large number of agencies. Fishermen are not very aware that there are other stages of cold water immersion that may pose a greater threat to their lives than hypothermia (stage 3). Cold shock (stage 1) and swimming failure (stage 2) appear to be responsible for more drowning fatalities in BC's commercial fisheries than either hypothermia or post-rescue collapse (stage 4). What is also little known is that being prepared to prevent or react to a cold water immersion event can greatly reduce the sometimes fatal effects of entering the water.

It is important for fishermen to become aware of the human body's physiological responses to the first two stages of cold water immersion. Vessel crew must understand how important keeping afloat in the water, which gives a rescue crew time to react, can be in improving their chances of survival. They must also understand and believe that how well they swim in warm water has no bearing on what they will be able to do in cold water. Only when fishermen understand what happens to the human body during the first two stages of cold water immersion will they willingly take the steps and precautions necessary to save lives.

The physiological effects of cold shock and swimming failure in themselves are not necessarily fatal. Hypothermia generally does not become a significant threat to survival for upwards of 30 minutes. It is therefore clear that keeping an individual's head above water and keeping the person afloat at the surface without requiring an expenditure of energy will substantially increase the chance of survival. Federal legislation in Canada requires that all vessels of a certain class, including many fishing vessels, carry one inherently buoyant keyhole style lifejacket for every person on board. However, the health and safety of crew on fishing vessels in Canada are the jurisdiction of the provinces and territories. In British Columbia, provincial regulations require that a personal floatation device or life jacket be worn whenever a work activity exposes a worker to the risk of drowning. The Workers Compensation Act allows various devices to be worn, including inherently buoyant lifejackets and manual and auto-inflating personal floatation devices. It should be noted that some fishermen contend that inherently buoyant life jackets are impractical for work.

Recently, a review of all occupational drownings reported to the Board from 1976 to 2002 was undertaken. This report ("A Retrospective Analysis of Drownings Reported to the British Columbia Worker's Compensation

Board 1976-2002," C.J. Brooks and K.A. Howard, unpublished) found no evidence that victims died from cold shock or swimming failure while wearing floatation devices. This determination, though, cannot be confirmed as many of the incident reports lacked sufficient detail to make this conclusion absolute.

Understanding, even at a basic level, the four stages of cold water immersion and their physiological effects should lead vessel masters and crew to being better prepared for immersion events before they happen. A three-point solution to address these issues is provided as follows:

1. Arrange the vessel's decks and work procedures to ensure crew exposure to entering the water is reduced.
2. Stay out of the water! If there is a risk of entering the water, be prepared to stay afloat to survive the effects of cold shock, to reduce the need to swim, and to give rescuers time to react.
3. Wear a personal flotation device or lifejacket if you are exposed to the risk of entering the water. A person who is unprotected by a floatation device can drown very quickly, often in as little as 3 minutes from initial immersion.
4. Develop and practice procedures to recover a crew member who is overboard quickly and be prepared to abandon a distressed vessel in a manner that eliminates or minimizes the need for crew to become wet!

There are many different types of lifejackets and personal flotation devices available on today's market. It is unlikely that any device will work well in all fisheries or even across all seasons and weather conditions. Auto-inflate flotation devices offer excellent floatation in a small, lightweight, unrestrictive package and are used successfully throughout the year in British Columbia. Immersion suits (survival suits) are required by provincial regulation for each crew member on every fishing vessel, but are far too bulky to work in. They do, however, save countless lives when there is enough warning to allow crew to get them on before they find themselves in the water. Between these ends of the equipment spectrum, there are many variations that can be explored.

A final recommendation is for the collection of much more data on the issues surrounding cold water immersion. As stated previously, a recent review of WCB files on drowning fatalities concluded that most investigations were incomplete when it came to determining the events that led to the drowning. It is important to understand that the preventative measures for reducing the risk of hypothermia are different from the preventative measures that need to be put in place to reduce the risks posed by cold shock and swimming failure. The WCB, in conjunction with Dr. Chris Brooks and his assistant, Kimberley Howard, has developed an investigation checklist for immersion-related fatalities. This checklist is found in the appendix to this report.

References

Brooks CJ (2003). *Survival in Cold Waters.* Marine Safety Directorate, Transport Canada.

Golden F and Tipton M (2002). *Essentials of Sea Survival.* Human Kinetics Europe, Ltd.

Appendix: Investigation Checklist–Water-Related Incidents

INVESTIGATION CHECKLIST - WATER RELATED INCIDENTS
(Fill in a separate form for each worker)

PART 1 – Physical Information

Category of Incident (Fishing, Logging, Diving, MVA, etc) _____

If Vessel involved   Name Of Vessel _____
    Length of Vessel_____
    Vessel Registration Number _____
    Vessel License Number _____

If Vehicle involved, type of vehicle (type, make and model) _____

Speed of vehicle or vessel at time of accident _____

Activity of vehicle or vessel at time of accident_____

Turnaround time of the vehicle/vessel_____

Estimated distance from shore or edge of river _____ (metres, kilometres)

Did incident occur in: ☐ ice   ☐ water

Depth of water _____(metres)

Time of day _____(24 hour clock)

Daylight: ☐   Twilight: ☐   Darkness: ☐   Other: ☐ _____

Weather conditions: Observed   Estimated
    Air Temp (°C) _____ _____
    Water Temp (°C) _____ _____
    Sea State (metres) _____ _____
    Wind Speed (knots) _____ _____
    Direction _____ _____

Brief description of the incident:

_____
_____
_____
_____
_____
_____
_____

PART 2 – Human Factors Information

Number of people involved _____

Number of persons injured _____

Number of persons fatally injured _____

Injured Workers Information

Name _____   Date of Birth _____

Height _____   Weight_____

☐ Fatal   ☐ Injury

Body Core Temperature at Site   ___°C   Body Core Temperature at Hospital ___°C

Could worker swim? ☐ well   ☐ average   ☐ poor   ☐ no   ☐ not known

Did the worker sustain injuries other than immersion injuries? ☐ yes   ☐ no   ☐ not known
    If yes, describe

    _____
    _____

Had worker taken any survival training?   ☐ yes   ☐ no
    If yes, explain

    _____
    _____

- Was worker alive when entered the water?   ☐ yes   ☐ no   ☐ unknown

Does autopsy clarify whether worker was alive or dead upon entry   ☐ yes   ☐ no

If the worker was dead before entering the water then this is not a water related incident and there is no need to proceed further. If it remains unclear as to the condition of the worker prior to entering the water then continue as if the worker were alive as s/he entered the water.

- If a motorized vehicle or piece of equipment was involved, was a seat belt available?
    ☐ yes   ☐ no

Was a seat belt used?   ☐ yes   ☐ no

- Was the worker physically trapped within the vessel, vehicle, etc.? ☐ yes   ☐ no
    Describe:
    _____
    _____

How easy or difficult is it to operate the doors/windows underwater? Describe.
    _____
    _____

Do post mortem reports aid in answering above questions?   ☐ yes   ☐ no ☐ undetermined

- If the reports do help, then was worker *physically* drowned (i.e. entrapped) or did they drown by other means?
    ☐ yes, physically drowned   ☐ no, drowned by other means

Was worker observed to: (tick appropriate box)
    ☐ make ineffective swimming strokes/struggle violently and appeared to be alive/conscious and even
    reaching for life rings etc?
    If so, for how many minutes? _____

☐ commence to swim either to other vessel or shore?
- If yes, how many minutes or hours? _____.
What distance was covered before worker seen to succumb?_____(metres/kilometres)

- Can you conclude how soon after water entry did worker succumb?
  - ☐ within first 5 minutes
  - ☐ between 5 and 30 minutes
  - ☐ after 30 minutes
  - ☐ at or shortly after rescue
  - ☐ unknown

- Length of time in water before retrieval?
  - ☐ under 5 minutes
  - ☐ between 5 and 30 minutes
  - ☐ over 30 minutes
  - ☐ over 1 hour
  - ☐ unknown

Were the above estimations determined from:
☐ witness testimony
☐ investigator's estimation (an estimate is encouraged if it is possible and there is no witness testimony)
☐ other (i.e. video) Describe_____

- How well clothed was the victim?
  - ☐ light ☐ medium ☐ heavy ☐ unknown

Briefly describe how worker was clothed:
_____
_____

Did worker wear a lifejacket, floater coat, pfd, etc.?
☐ yes   ☐ no ☐ unknown.

Describe type, make, age and model:
_____
_____

- Did floatation device perform to specifications? ☐ yes   ☐ no   ☐ unknown

If 'no', describe what went wrong:
☐ unable to put on while in the water
☐ came off in the water
☐ not secured correctly
☐ poor maintenance
☐ inappropriate selection of gear
☐ did not inflate
☐ punctured
☐ damaged
☐ did not provide enough buoyancy
☐ other, describe:
_____
_____

- Were PFD's/Lifejackets carried on vessel or in vehicle? ☐ yes   ☐ no   ☐ unknown

- Were there enough PFD's/Lifejackets for all on board? ☐ yes ☐ no ☐ unknown

Did worker wear an immersion suit? ☐ yes ☐ no ☐ unknown
Briefly describe the type (make and model number):
_____
_____

- Were immersion suits carried on the vessel? ☐ yes ☐ no ☐ unknown

- Were there enough immersion suits for all on board? ☐ yes ☐ no ☐ unknown

Did immersion suit perform to specifications? ☐ yes ☐ no ☐ unknown

- If 'no', describe what went wrong:
  - ☐ sinking occurred too quickly to locate and don.
  - ☐ stowed in a place already underwater
  - ☐ physically stowed in an inaccessible place
  - ☐ no knowledge that suits were on board
  - ☐ suits leaked badly
  - ☐ no training in how to don/doff the suits
  - ☐ poor maintenance
  ☐ other, describe:
  _____
  _____

- Were lifeboats, life rafts or skiffs carried? ☐ yes ☐ no ☐ unknown
  - What type? _____

Were lifeboats, life rafts or skiffs deployed? ☐ yes ☐ no ☐ unknown

- If deployed, did they perform to specifications? ☐ yes ☐ no ☐ unknown

 If 'no', describe what went wrong:
With Launch
  - ☐ in ability to launch
  - ☐ vessel sank too quickly to launch
  - ☐ didn't know how to launch
  - ☐ too much list to launch
  - ☐ already underwater
  - ☐ weather made it too difficult to launch
  - ☐ got entangled
  - ☐ blew up against side of sinking vessel
  - ☐ failed to inflate
  - ☐ punctured
  - ☐ crew too cold and fatigued to launch
  - ☐ other, describe:
  _____
  _____

When in Water
- ☐ capsized
- ☐ blew over
- ☐ deflated
- ☐ washed overboard
- ☐ inability to get back in after being washed overboard
- ☐ flooded
- ☐ other, describe:

_____

_____

- How many hours/days did the worker work in the last:
    - 24 hours _____ hrs
  last one week   _____ days
    - last one month   _____ days

Was there any medical condition that contributed to the incident (i.e. Epilepsy, Diabetes, Heart Disease, etc.)
☐ yes   ☐ no   ☐ unknown
If 'yes', describe: _____

- Were any other medication, drugs or alcohol involved? ☐ yes   ☐ no   ☐ unknown
  (Ask specifically about antihistamines and anti-fungals and whether medication was prescribed or over-the-counter)
  Describe: _____

Include a brief description of the autopsy results:

_____

_____

_____

_____

- State the Coroners Cause of Death:

_____

_____

_____

Final determination
- ☐ Drowning from entrapment
- ☐ Drowning from Cold Shock
- ☐ Drowning from Swimming Failure
- ☐ Drowning or Death from Hypothermia
- ☐ Drowning or Death from Post Rescue Collapse
- ☐ Drowning or Water Related Death from Undetermined Cause
- ☐ Drowning – Other _____

Checklist for water related fatalities: C.J. Brooks, K.A. Howard, and S.K. Neifer

SESSION ELEVEN:
REGIONAL APPROACHES TO FISHING SAFETY, II

Port of Valdez, Alaska (Photo courtesy of Alan Sorum)

SAFETY AND SURVIVAL OF FISH WORKERS FOR MULTI-DAY FISHING CRAFT IN SRI LANKA

Herman Kumara
Fisheries Solidarity
No. 10 Malwatta Road
Negombo, Sri Lanka
E-mail : nafsoone@slt.lk / fishmove@slt.lk

Abstract

This presentation discusses the role of policy change and injury surveillance to enhance work-related safety for commercial fishermen in Sri Lanka, whom we shall refer to using a local acronym, MDB, for groups of fish workers who use multi-day boats. Historically, the safety of commercial fishermen has not had a high priority in terms of policy enforcement. Poor fish workers are routinely hired for temporary work aboard vessels, and vessel owners do not consider the safety of these itinerant workers as a priority. The injury risks to these workers are very high, as these fishing vessels are at sea for longer periods, operating with very poor working facilities. Vessels often are missing first-aid boxes, and seldom would one find any medical supplies to treat workers. Safety equipment is usually missing as well. In addition to the lack of safety equipment and first-aid treatment aboard Sri Lanka's commercial fishing vessels, there is no proper reporting mechanism to notify authorities when injuries or fatalities occur. This presentation describes the need for on-going surveillance for commercial fishermen in Sri Lanka and then discusses the possible parties that could help create safer working conditions for fishermen there.

Introduction

Sri Lanka is an island situated in the Indian Ocean with a 1,775-km-long coastline. All of Sri Lanka's populace lives within 100 km of the coastline. Fishing is very important to Sri Lanka's economy, as it provides employment, nutrition, and food security to the nation. Recently, the foreign exchange earnings from fish products have grown in importance to the national economy. According to the reports of Ministry of Fisheries and Ocean Resources (MOFOR), the major contribution from marine fisheries production comes from the coastal fisheries. Deep sea commercial fisheries employing multi-day boats (MDBs) comprise around 25% of the total marine fish production of the country. Fisheries in offshore and deep-sea waters are composed primarily of medium-sized and large pelagic species, mainly tuna, and an exotic range of deep sea demersal species. Yellow-fin tuna, big-eye tuna, and skipjack tuna, all of which are highly migratory species, as well as sharks, are the major targets for MDBs. Around 1,500 MDBs work from Sri Lanka. A marked increase in fish production from offshore or deep-sea areas has occurred owing to the introduction of better-equipped MDBs. Production from the offshore and deep-sea sector increased from 800 metric tonnes in 1972 to 84,400 metric tonnes in 2000. The income distribution pattern of the small-scale and commercial fisheries in the country is shown in Table 1.

Table 1: Multi-day craft in Sri Lanka

Craft	Production, kg	Revenue, SLR
Multi-day boat	35,164	2,250,500
FRP boat	9,580	479,000
Motorized traditional craft	7,100	355,000
Nonmotorized traditional craft	800	41,000
Nonmotorized teppam	2,000	100,000
Source: Comparative study on the economics of large- and small-scale fishing operations in Sri Lanka. Working paper NARA/SED/02/1998		

Annual fish production and average revenue of fishing vessels

The average net income of an MDB crew is higher than the income of other boat owners. The current value for an MDB is around SLR 4 million. (1 US $ = 94 SLR)

Average cost of multi-day craft

Comparative studies were undertaken at two major fishing harbours on the west and south coasts. Labour and other expenses for MDBs (Table 2) are rather higher than those associated with small-scale and artisanal fisheries in Sri Lanka. Table 3 shows average annual variable costs of craft operations of MDBs.

Table 2: Average cost of multi-day craft

Item	32 to 34 ft	34 to 36 ft	36 to 39 ft	Over 40 ft
Hull	475,000	650,000	750,000	1,550,000
Engine	400,000	450,000	450,000	1,200,000
Gear/accessories	425,000	520,000	700,000	1,250,000
Total	1,300,000	1,620,000	1,900,000	4,000,000

Source: Economic and Social implications of multi-day fishing in Sri Lanka. O. Amarasinghe, Oct. 2001.

Table 3: Average annual variable costs of multi-day craft

Item	32 to 34 ft	34 to 36 ft	36 to 39 ft	Over 40 ft
Labour	1,300,346	941,466	894,144	1,837,506
Fuel	521,854	455,143	621,450	720,652
Other inputs*	809,734	816,928	674,623	923, 752
Total	2,631,933	2,213,537	2,190,217	3,481,910

* Includes food, water, ice, maintenance, license fees, handling charges, payments for watchers at anchorage, cost of cleaning and loading, gate charges. etc.

Drift gill nets and longlines are the main types of gear used in Sri Lankan fishing. A few troll lines are used. Generally, gear on MDBs consists of 50 pieces of drill gill nets, 100 longline baskets, and five to ten troll lines. Most of MDBs are equipped with navigational instruments such as radio transmitters and satellite navigators. Some of the MDBs have winches for net hauling.

Most MDBs sail more than 36 hours before actual fishing starts. Generally a fishing trip continues for 10 to 45 sea days, but 15- to 25-day trips are not unusual. It is important to note that the MDB owners are getting comparatively higher returns for their investment, but appear to pay inadequate attention to the safety of the fish workers. As stated earlier, safety of fish workers is not considered a high priority, possibly because most of the boat owners do not accompany fishermen to the sea. Owners employ temporary fish workers on their boats, usually managed by a skipper who is either a friend or a relative of the boat's owner. Employer–employee relations have undergone many changes with market expansion, population growth, and the advent of new technology. The crew is no longer employed on a permanent or long-term basis and move from one employer to another. Temporary job assignments for fish workers (who may be hired for only one fishing trip) weaken their bargaining power to improve working and safety conditions aboard the Sri Lankan fishing fleet (Table 4).

Table 4: Crew data

Catchment area	Classification by length	Crew compliment	Competency certificate holders
Coastal, offshore	Small fishing vessel, less than 12 m long	3-4	None
Offshore, deep sea	12-24 m long, propulsion less than 750 kW	Crew: 5; skipper: 1	None
Deep sea, high seas	24+ m long, propulsion 750 kW or more	Crew: 5; skipper: 1	None
Fishermen's log			None
Maintain log			None

Owners of MDBs expect the crew to work at sea until they catch the maximum amount of fish. Boat owners keep in touch with their boats through radio messages, informing the boat when to come back to shore in order to benefit from maximum income from the fish catch.

Some sea safety issues related to multi-day fisheries industry

An analysis of engines used in MDBs showed that 40% were reconditioned and had been chosen for economic reasons. But the durability of reconditioned engines is poor at times and can create risks for fishermen. Currently, no policy addresses the need for reliable engine power in the Sri Lankan fishing fleet. No Sri Lanka regulations exist to assess and document competencies for commercial fishermen. (Note the lack of certified commercial fishermen in Table 4.)

Sri Lankan fishermen often face unpredictable weather, unstable surfaces, and mechanical and flammable hazards. It is very important to report that the Sri Lankan MDBs operate all over the Indian Ocean, spending very long periods at sea and working in poorly equipped boats (Table 5). The lack of concern for fishermen's welfare forms a strong concern in fish worker organizations in Sri Lanka. Many Sri Lankan fishermen feel compelled to cross national borders to look for better working conditions. The situation is exacerbated when there is competition among commercial fishing fleets owned by foreign companies, which offer boats with sophisticated equipment, and MDBs operated by local fish workers, which offer very little in the way of amenities.

Table 5: Areas of operation and duration of voyages

	32 to 34 ft	34 to 36 ft	36 to 39 ft	Over 40 ft
Area of operation	EEZ	EEZ and international waters	EEZ and international waters	EEZ, internal waters, and territorial waters*
Duration of trip	1 week	1-3 weeks	1-3 weeks	3 weeks to 3 months
*Territorial waters include Andaman Islands, Nicobar Islands, Maldives, Australian Island, Bangladesh, Thailand, and Madagascar.				

Modern coastal fishing craft in Sri Lanka have been undergoing drastic changes in terms of construction materials, size and shape of hulls, deck layouts, and navigational accessories, but all these improvements have been introduced in piecemeal fashion. Stricter requirements should be implemented to improve minimum standards for working facilities on MDBs through new regulations attached to an MDB license. National requirements would go a long way to improve working conditions for Sri Lankan fishermen. Recently MOFOR provided first-aid boxes to many MDBs that lacked this box, which is a very important piece of equipment on any commercial fishing vessel. As we assessed safety equipment aboard some of the MDBs we encountered, many of the crew said that there was a first-aid box on their vessel but no medicine. Furthermore, the crew had no idea how to identify or use what little medicine might be found. MOFOR thus suggests that there should be on-going training to educate fish workers on health, first aid, sea safety, and survival.

Need for standard safety devices and mechanisms

Sri Lankan fishermen have said that most of the standard safety devices are not affordable for the crews of fishing vessels. MOFOR notes that in spot checks, it has seldom found safety devices aboard fishing vessels. Absence of effective search-and-rescue or any other emergency equipment has also been noted; the lack of this equipment increases risks of danger at sea. Sri Lanka lacks the capacity for swift rescue of stranded fishing vessels. Thus, it is important for Sri Lanka to develop a systematic, shore-based system to log departure times, estimate time or dates of arrival, and possible destinations against vessel numbers to facilitate search-and-rescue missions.

Rough weather conditions at sea

Since Sri Lanka is located in the equatorial region, its fishing grounds spread across a number of major surface current systems. Moreover, these fishing grounds are strongly influenced by northeast-southwest monsoon weather conditions, with very strong winds that blow in opposite directions. The number of fishing boats drifting in the open sea because of engine failure, navigational difficulties, or fuel shortage remains high, particularly among MDBs, and each of these drifting vessels is at the mercy of harsh weather. Drifting and lost MDBs compose a major problem for the Sri Lankan government. Some boats have been larger ocean-going vessels sailing in the area.

Threat of capture and detainment in neighbouring countries

Another major safety issue at present that impacts Sri Lankan fishermen is the threat of capture, arrest, or conflict at sea. The number of fishing craft captured and the number of Sri Lankan crew members detained in foreign jails has increased drastically during the past years. Some Sri Lankan MDBs have been captured by various countries in the neighbouring Indian Ocean. The civil war in the north and east of Sri Lanka has also created threats to the lives of the fish workers on board both artisanal craft and MDBs. The problems associated with civil unrest are now 20 years old and continue to affect the safety of Sri Lankan fishermen.

Conclusion

It is very important to the Sri Lankan fishing community that the various UN agencies have taken important steps to improve working conditions and safety at sea for commercial fishermen around the world. As national governments ratify these policies, concrete measures can be taken to improve safety for fishermen. We especially acknowledge the importance of the following:

- Code of Conduct for Responsible Fisheries, by FAO in 1995.
- ILO Instruments on Fisheries Sector. Five conventions and two recommendations apply to persons working on fishing vessels. The existing conventions concern minimum age, medical examinations, articles of agreement, accommodations, competency certificates, vocational training, and hours of work.
- Bay of Bengal Programme's Chennai declaration.

Although it is focused on small-scale and artisanal fishermen's sea safety, the Chennai declaration is very important for addressing safety issues for sea-going fishers.

References

Amarasinghe O (2001). Economic and social implications of multi-day fishing in Sri Lanka.

Ministry of Fisheries and Ocean Resources (1998). Comparative study on the economics of large- and small-scale fishing operations in Sri Lanka. Working paper NARA/SED/02.

SRI LANKAN FISHING INDUSTRY
AND SAFETY AT SEA

G. Piyasena
Director General
Department of Fisheries and Aquatic Resources
Sri Lanka

Abstract

Commercial fishing is one of the most important sectors in the economy of
the island of Sri Lanka, providing direct and indirect employment to around
250,000 persons. The sector also contributes nearly 3% to the Gross Domestic Product of the country and provides the bulk of the animal protein
consumed by the population. During the past two decades, fishing activities
have been extended from the area within the continental shelf up to the edge
of the 200-mile Exclusive Economic Zone (EEZ) and beyond. The mechanization programme has resulted in a motorized fishing fleet of nearly 14,000
vessels. The larger boats operate on a multi-day basis and fish in offshore
areas. Although Sri Lanka has a 200-mile EEZ, its full potential is yet to
be realized. In the meantime, frequent reports are heard that foreign vessels
enter the EEZ, carry out illegal fishing, and damage fishing nets of the local
boats. The present Coast Guard service is unable to carry out any effective
monitoring, control, and surveillance (MCS), and Sri Lanka is now in the
process of attempting to build a cost-effective MCS system for this purpose.

Introduction

The island Republic of Sri Lanka is located in the Indian Ocean southeast
of India between 5°55′ and 9°51′N latitude and 79°4′ and 81°53′E longitude.
Its area is approximately 65,610 sq km, with a coastline of about 1,620 km,
Sri Lanka and the southern tip of India stand on the same continental shelf
and are separated by the shallow Palk Strait, which is barely 30 metres deep.

Sri Lanka's Maritime Zones Law No. 22 of 1976, proclaims several areas of national maritime jurisdiction in conformity with the provisions of the UN Convention on the Law of the Sea, namely "Internal Waters," "Historic Waters," "Contiguous Zone," and the "Exclusive Economic Zone." The area enclosed by the Exclusive Economic Zone (EEZ) is reported as 517,000 sq km, 7.8 times the total land area of the country.

Over the last five decades, the fisheries sector of Sri Lanka has undergone a significant transformation, resulting in the modernization of an artisan fishery that was previously exclusively based on the use of rudimentary craft, fishing methods, and gear. While the fishing fleet of the early 1950s was made up only of traditional craft, such as dugout canoes (*oru, wallam*) and log craft (*teppam, kattumaram*), by the year 2002, the number of traditional boats totaled 16,376, or only 58% of the 28,135 vessels making up the entire fleet. The remaining 42%, numbering 11,759, are modern fishing boats made out of timber or fibre-reinforced plastic and powered by inboard and outboard engines.

Concomitantly, there has also been a similar change in the types of fishing nets and gear used. In the early 1950s, the most important fishing method was beach seining (*madel*), reported to have accounted for almost 40% of the total marine fish landings. Today, however, beach seining has paled in significance, and its contribution to overall production is rather marginal. Gill netting is now the most important fishing method in terms of the contribution made to production. In addition, the nets and gear that were made locally of vegetable fibres have given way to those made out of such synthetic fibres as nylon and kuralon. These changes were brought about through a massive extension effort launched by the Department of Fisheries with the technical and financial assistance of the Food and Agriculture Organisation and several donor countries and organizations. As a result of this modernization, landings of marine fish have increased from 51,000 metric tons in 1960 to 303,000 metric tons in 2002, nearly a six-fold increase.

In addition, gradual growth in inland fisheries production in Sri Lanka reached its climax in the late 1980s. This was brought about by a programme of stocking of large inland water bodies with fish and was supplemented by an increased harvesting effort made possible through the issue of subsidized craft and gear among the communities near these stocked bodies of water. The overall result was a very substantial increase in national fish landings, from around 1,500 metric tons in 1960 to 39,720 metric tons by 1989, an increase of nearly 26 times within just three decades. However, due to a fundamental change in government policy in the late 1990s that resulted in the withdrawal of government support for inland fisheries, a downward trend in inland fish production is seen from 1990 to 1998. Thereafter, inland fish production picked up once again, and by 2000, it had reached 28,000 metric tons.

Due to these impressive increases in marine and inland fish production, it has been possible to increase the per capita availability of fish from a level of 11 kilograms in 1978 to 18.5 kilograms in 2002. The rural population has benefited considerably in terms of better availability of fish due both to better transport of marine fish into the hinterlands and to increased production of freshwater fish.

Fisheries sector in the national economy

Fishing has been a major economic activity on the island from time immemorial, and it has been the main livelihood of large numbers of coastal dwellers. Since Sri Lanka's political independence in 1948, successive governments have increasingly devoted attention to fisheries development and have carried out many programmes for the advancement of the industry. A significant characteristic of the fishing industry in Sri Lanka is that it has always been dominated by the private sector, although not the formal private sector. Except for a handful of boats owned by a few fisheries' co-operative societies or by a few companies, the rest of the fishing boats and gear deployed in the industry are owned and operated by thousands of individual fishermen, family units, or informal groups. The role of the government in these circumstances has been one of facilitation, promotion, encouragement, and regulation of activities in the national interest.

The fishing industry plays a major role in providing animal-based protein, an important element of the diet of the Sri Lankan population. According to the Food Balance Sheets from the Department of Census and Statistics, fish has consistently contributed around 65% of the animal-based protein intake of the population. In addition, it is also noteworthy that in Sri Lanka, with a multireligious and multiethnic society, numerous religious and cultural biases and prejudices prevail against the consumption of animal flesh. However, fish is the item preferred by most and hence is always in great demand.

Fishing has been the most important economic activity in the coastal areas of the country, and it is estimated that, at present, nearly 150,000 persons are directly employed in the fishing industry, including inland fishing. An additional 100,000 persons are estimated to be employed in several fishery-related economic activities, such as boat building; fish net manufacturing; ice production; processing, trading and marketing of fish; and in providing other services required by the industry, such as transport, repair of engines, and hull work. It is also estimated that there are nearly 1,250,000 persons, including the dependents of the industry participants, who derive their sustenance from the industry. Its contribution to the Gross National Product has stood around 2% in the past few years.

In recent years, the fisheries sector has also emerged as an important source of foreign exchange through the exports of several items of high-value fish and fishery products, including chilled and frozen tuna, shrimp, lobsters, shark fins, and sea cucumbers. Starting from humble beginnings in the late 1970s, the total value of fishery-based exports has been continuously on the increase, reaching a level of Rs. 8,047 million (US $83 million) in 2002.

Sri Lanka's coastal fishery resources are made up of the exploitable pelagic and demersal marine species of the entire water column on the continental shelf. About 70% of the exploitable coastal resources consist of small pelagics, such as sardines, herring, anchovies, mackerel, and flying fish. Oceanic large pelagics, such as tuna, marlin, shark, sailfish, and swordfish, are also caught in the coastal waters of the country. The common oceanic pelagics are the yellow-fin tuna (*Thunnus albacares*), skipjack (*Katsowonus pelamis*), Kawa-Kawa (*Euthyus affinis*), frigate tuna (*Auxis spp*), and seer fish (*Scomberomorus commersoni*). Demersal species caught include emperors, snappers, groupers, sweetlips, sciaenids, carangids, breams, goatfishes, and leiognathids, as well as invertebrates such as squid, prawns, crabs, and lobsters. The commercially important food fishes caught in the coral reefs are the groupers, snappers, emperor fishes, rabbitfishes, sweetlips, surgeonfishes, parrotfishes, and barracudas. Important invertebrates caught are spiny lobsters (*Panulirus versicolor and P. Ornatus*), octopus, sea cucumbers, squid, and cuttlefish. Many other species are exploited for the ornamental fish industry; these include butterfly fish, surgeon fish, blennies, dragonets, gobies, wrasses, file fish, angelfish, and damselflies, among others.

Since production from the coastal sector has almost reached optimum levels, any further increase in marine fish production is largely dependent on the exploitation of resources in offshore areas. The offshore fishery targeting migratory stocks of tuna, billfish, and other deep water species was the fastest growing sector in the marine fishing industry during the last decade.

Catch and effort

Sri Lanka's current marine fishing fleet consists of a little over 28,000 craft of several types. Despite the implementation of a series of development programmes for the modernization of the industry, supported by technical and financial assistance from the international community, Sri Lanka's fishing fleet still retains much of its artisanal and traditional character. Even now, 59% of the total fishing craft of Sri Lanka are of the indigenous, traditional type. The most important development in the fishing industry during the past two decades was that fishing activities by the local fishing fleet have extended into areas outside the continental shelf. This was made possible by the design and the introduction of new types of fishing boats. These newer, larger boats have the capability of fishing continuously over several days, primarily due to the installation of an insulated fish hold, better facilities for the crew, and the capacity to take more fuel and water. These boats, called multi-day boats, have enabled the country to harvest commercially valuable large tuna and other resources found in the outer fringes of the EEZ of Sri Lanka. It is also noteworthy that, at least in the initial stages of its development, multi-day fishing activities were actively promoted by the government through schemes of subsidies and credit just as much as the initial craft mechanization programmes were promoted through government efforts.

With regard to fishing methods and gear, the most noteworthy feature is that gill netting is the most important fishing method in terms of production. In the coastal fisheries, small-mesh gill nets are used in the exploitation of small pelagic fish. This method accounts for around 80% of the landings of these varieties. Beach seines account for the balance. Gill nets of 60- to 80-millimetre mesh size are the main gear utilized for the medium-sized pelagics, such as small tuna varieties and Indian mackerel. The large gill net is used predominantly in taking larger pelagics, including skipjack, yellow-fin tuna, and Spanish mackerel. Longlining is the next most important gear and takes two forms, tuna longlining and shark longlining.

As already mentioned, marine fish production has recorded a steady overall increase during the past 25 years, except for a decline in 1984 due to the civil disturbance in the northern and eastern parts of the country. Coastal fish production showed a 0.9% increase per annum between 1978 and 1986, and then declined to an annual growth rate of 0.23% between 1987 and 1997. Production from this subsector peaked in 2002 and reached 176,250 metric tons in that year. There has also been a continuous increase in offshore fish production in the recent past. Overall, the increase in production in the 1990s could be attributed to the introduction of more efficient engines and more reliable fishing craft, enabling the expansion of fishing from the traditional fishing grounds in lagoons, estuaries, and inshore waters toward the oceanic waters offshore. Table 1 shows the past trends in marine fish production in Sri Lanka.

More than 15 fishery resource surveys have been conducted in Sri Lanka since 1920, mostly on pelagic resources. Because the surveys were not followed by reliable statistical data collection, it is not possible to conclude if the maximum sustainable yield has already been attained. There are indications, however, that, with the exception of a few fish stocks, the harvest in the country's coastal waters has already reached its optimum level, and some stocks have already manifested signs of overexploitation. Small-scale fishers report decreasing sizes of fish and lower catch volumes with greater fishing effort. In this light, the Ministry has adopted a "precautionary approach" in regard to its development policies.

Table 1: Fish production in Sri Lanka

Year	Coastal fish production	Offshore fish production	Total marine fish production
1997	152,750	62,000	214,750
1998	166,700	73,250	239,952
1999	171,950	76,500	248,450
2000	179,280	84,400	263,680
2001	167,530	87,360	254,890
2002	176,250	98,510	274,760

Policy framework

No clearly articulated and separately documented set of policies pertaining to fisheries and aquaculture was produced until 2002. However, the evolution of policy thrust areas could be identified in terms of fisheries development plans formulated between 1959 and 2002. There have been several planning cycles, starting with the 10-year plan of 1950 formulated by the National Planning Council. The five-year plan of 1972-1976, formulated by the Ministry of Planning and Employemnet, followed it. Subsequently, four more development plans were formulated by the Ministry of Fisheries.

- Fisheries Master Plan, 1979-1973
- National Fisheries Development Plan, 1990-1994
- National Fisheries Development Plan, 1995-2000
- Six-Year Fisheries Development Programme, 1999-2004.

Current fisheries policy is clearly laid down in the National Fisheries Policy and the Development Plan, formulated in 2002 by the Ministry of Fisheries and Ocean Resources (MFOR). The main elements of the policy are stated below.

1. MFOR is committed to performing the principles of "responsible fishing" and will create the necessary awareness and management programs to achieve this purpose. Stakeholder involvement in the management of fisheries is a key policy objective.
2. A reliable database producing meaningful information to support the applications of proper fisheries management initiatives will be developed.
3. The Fisheries and Aquatic Resources Act and other laws and regulations made thereunder will be revised to facilitate effective fisheries management and community participation to prevent harmful fishing practices and to ensure the conservation of resources and sustainable fisheries.
4. Coastal, lagoon, and inland fisheries will be developed and sustained primarily to provide the communities associated with them with nutritious food security, livelihoods, and income-earning opportunities. MFOR will, through its agencies, provide or facilitate the delivery of requisite inputs, including management systems, that enable the above purposes in an equitable and sustainable manner.
5. A high priority will be accorded to the conduct of resources surveys, stock assessments, and exploratory fishing to build the information base relating to fish resources in costal, offshore, and deep sea areas.
6. MFOR will take measures to improve productivity in the fishing industry through the introduction of appropriate and advanced technologies in catching, processing, and marketing fish and fishery products.
7. MFOR will prevent the use of fishing practices that are destructive to the resource and fish habitat, particularly the use of dynamite and stupefying substances, through awareness programs, effective surveillance, and stringent enforcement of laws. The existing laws will be revised to deter such activities.

8. MFOR will also actively support the diversification of fishing methods to lessen the reliance on gill netting, as well as to lessen the incidence of resource- and environment-damaging fishing practices, and to promote the harvesting of underexploited and unexploited resources. A commission headed by a retired judge, with powers to call for public hearings and expert evidence and to consult communities, will be appointed to determine the fishing methods that should be disallowed.

Legal Framework

The Fisheries and Aquatic Recourses Act No. 2 of 1996 provides the legislative framework for development and management in Sri Lanka. The act came into operation in 1996 and replaced the fisheries ordinance of 1940, which had been amended several times between 1950 and 1979. The current act provides for the management, regulation, conservation, and development of fisheries and aquatic resources of Sri Lanka. Furthermore, it provides for the following:

+ Appointment of offices, including authorized officers and their powers.
+ Establishment of a fisheries and aquatic resources advisory council.
+ Preparation of a plan for the management, regulation, conservation, and development of fisheries and aquatic resources.
+ Licensing of fishing operations.
+ Registration of local fishing boats.
+ Protection of fish and other aquatic resources.
+ Conservation.
+ Aquaculture.
+ Citations and penalties.

It is proposed to replace the current Fisheries and Aquatic Resources Act by a new act formulated by foreign and local legal experts with the assistance of the Asian Development Bank to ensure that the national fisheries policy can be implemented effectively and that management and development objectives are achieved. The key features of the draft fisheries management and development bill are to—

+ Provide a comprehensive, integrated framework for managing and developing the fisheries resources for the benefit of the people of Sri Lanka.

+ Include institutional reforms initiated by this government to support the management and development functions and enable participatory management.
+ Implement Sri Lanka's international obligations to provide a clear basis for foreign fishing and related activities in Sri Lanka waters.
+ Cover a range of activities, including fishing, landing of fish by foreign fishing vessels, operating fish processing facilities, commercial test fishing, and manufacturing fishing boats and fishing gear.
+ Provide a framework for a clear, transparent licensing and registration system to ensure that the fisheries resources will continue to yield maximum benefits to the people of Sri Lanka.
+ Incorporate clear powers of officers and inspectors for monitoring controlled surveillance, including the use of cameras, position-fixing instruments, and other up-to-date technologies.
+ Provide an innovative and straightforward dispute resolution process.
+ Clarify the roles of the courts and legal procedures to enable clear and efficient enforcement.

THE ALASKA FISHERMEN'S FUND

James Herbert
Alaska Vocational Technical Center
Seward, Alaska, USA

K. Renee Cox
Alaska Department of Labor and Work Force Development
Juneau, Alaska, USA

Introduction

Commercial fishermen and women face a variety of hazards in their work. The financial consequences of injuries or medical conditions that arise on the job are not always met by the Social Security network. The State of Alaska is unique in the nation by providing its commercial fishermen and women a program that offers some relief and security if they are injured or become ill while engaged in their jobs. Under federal laws such as the Jones Act, owners of documented vessels have specific obligations with respect to injured crew members. On undocumented vessels, these regulations do not apply to the vessel owner or to the crew.

Purpose

The State of Alaska, through its Department of Labor and Workforce Development, administers a program that provides for the treatment and care of Alaska licensed commercial fishermen who have been injured while fishing in the state. The Alaska Territorial Legislature established the Fishermen's Fund in 1951 to help meet the medical needs of men and women in the commercial fishing industry. When Alaska became a state in 1959, the Fishermen's Fund was incorporated into statute (State of Alaska, Alaska Administrative Code [ACC], Title 23), and its operations and limitations described under the ACC (State of Alaska, AAC, Chapter 55). Because this dedicated Fund predates statehood, it has unique status compared with the financing of most Alaskan government entities. This paper will describe the Alaska Fishermen's Fund program and cite the benefits it has provided.

Methods and Description

Benefits from the Fund are financed from revenue received from each resident and nonresident commercial fisherman's license and fishery permit fee. Table 1 shows how receipts were generated for fiscal year 2003 (July 1, 2002-June 30, 2003) from the sales of licenses and permits. At the current time, 39%, or $23.40, of the resident crew license fee ($60) and $50 of a nonresident crew license fee ($180) are deposited in the Fund. Similar amounts come out of fishery permit fees. Since these amounts are collected by the Alaska Department of Fish and Game, there is an interagency relationship at work.

Table 1: Receipts for Fund, fiscal year 2003

Licenses sold during FY03				
Resident	9,393	39% X	$60 license fee	$219,796
Nonresident	7,061	$50 X	$180 license fee	$353,050
Duplicate	571	39% X	$5 license fee	$1,113
Child	348	39% X	$5 liense fee	$1,740
Nonresident child	48	39% X	$125 license fee	$2,340
Total	17,421			
Permits issued during FY03				
Resident	9,892	X 23.40		$231,473
Resident poverty		X 23.40		
Nonresident	3,486	X 50		$174,300
Nonresident poverty		X 50		
Total	13,378			
Total receipts				**$983,812**

The balance in the Fishermen's Fund has varied over the years, depending on the relationship of receipts, benefits paid, and administrative expenses. (For the period July 1, 1989–June 30, 2003 [FY1990-2003] see Tables 2 and 3.) The Fund is self-sustaining and currently has a positive balance of over $11 million. It is interesting to note that with the recent downturn in many aspects of the Alaska commercial fishing industry, there has been a drop in the number of licenses and permits sold over the past 14 years (Table 4). As a result, the Fund anticipates reduced revenues in the future. With fewer participants, there could be a drop in the total number of claims as well.

According to Alaska statutes, except for compelling reasons, the total benefit allowed for any one injury or accident is $2,500. Fishermen's Fund is considered an emergency fund of last resort, which means that benefits are awarded

Table 2: Overview of receipts and number of claims

Year	Receipts	Funds available at year's end	No. of claims
FY90	1,545,300	3,382,500	2,101
FY91	1,651,400	3,381,900	1,943
FY92	1,599,300	4,891,200	1,787
FY93	1,492,400	5,808,500	1,538
FY94	1,494,600	6,580,500	1,405
FY95	1,272,500	7,367,700	1,237
FY96	1,556,700	8,305,400	1,002
FY97	1,232,900	9,484,800	946
FY98	1,146,900	9,657,834	816
FY99	1,196,703	10,495,242	806
FY00	1,196, 999	10,729,131	856
FY01	1,182,554	11,447,962	786
FY02	1,079,755	11,717,248	808
FY03	983,946	11,815,543	696

Table 2: Overview of benefits paid and expenses

Year	Benefits paid	Administrative expenses	Total expenses	Revenue
FY90	644,400	163,300	807,700	1,545,300
FY91	593,300	168,000	761,300	1,651,400
FY92	593,900	172,900	766,800	1,599,300
FY93	493,400	243,700	737,100	1,492,400
FY94	445,700	222,900	668,600	1,494,600
FY95	339,400	206,000	545,400	1,272,500
FY96	290,200	187,100	477,300	1,556,700
FY97	447,500	188,500	636,000	1,232,900
FY98	399,967	198,982	598,949	1,146,900
FY99	597,542	211,576	809,118	1,196,703
FY00	497,998	201,748	699,746	1,196,999
FY01	531,366	211,964	743,330	1,182,554
FY02	584,408	223,565	807,973	1,079,755
FY03	584,408	223,565	807,973	983,946

only after full consideration of other coverage from private health or vessel insurance and public programs such as Veteran's Affairs or Medicare. The only exception to this payer of last resort is the Medicaid program, where the Fund pays first. Generally there is a 1-year limit from the date of claim approval for the claimant to complete his or her part of the process.

Table 4: Licenses and permits issued by Alaska

Year	Licenses	Permits*	Total
2003	17,422	20,002	37,424
2002	17,574	21,625	39,199
2001	20,653	23,456	44,109
2000	24,290	24,660	48,950
1999	25,060	25,292	50,352
1998	24,916	26,214	51,130
1997	27,238	27,512	54,750
1996	28,752	27,452	56,204
1995	31,391	27,731	59,122
1994	32,310	29,150	61,460
1993	33,011	29,861	62,872
1992	36,166	32,588	68,754
1991	36,390	32,798	69,188
1990	36,992	32,508	69,500
Total			773,014

*Active issued and renewed permits (not people) as of August 14, 2003.

The organizational structure of the Fund is straightforward. The Commissioner of Labor and Workforce Development oversees administration of the program with the assistance of the five members of the Fishermen's Fund Advisory and Appeals Council. The council is composed of the commissioner or his designee who serves as chairperson and five representatives from the fishing industry appointed by the governor for 5-year terms. The Council meets twice a year to review questioned claims, consider extensions of benefits and time, advise on policy, and conduct other business.

The day-to-day operations occur in the state capital in Juneau, Alaska, and are overseen by the fund administrator. This person writes and administers the budget, sets policy, prepares reports for the legislature, troubleshoots claims, and communicates with care providers and industry through an awareness program. The administrator supervises two Workers' Compensation technicians who process and evaluate claims, answer phone and e-mail inquiries, and pay bills. The council and staff seek to serve injured commercial fishermen effectively, to process claims quickly and efficiently, and to pay care providers in a timely fashion.

To qualify for benefits, crew members or skippers with an injury or illness directly connected to operations as a commercial fisherman must hold a valid commercial fishing license or limited entry permit issued by the state of

Alaska before the time of injury or illness. The onset of the injury or illness must be onshore in Alaska or on Alaska waters. Initial treatment must be received within 60 days of the onset of the injury or illness. An application must be submitted within 1 year after the initial treatment. Each treatment must be documented by submission of a physician's or physician's assistant's medical chart notes. It should be noted that persons holding a recreational fishing license and working on charter fishing boats are not covered. Similarly a worker who is directly connected to a seafood processing operation does not qualify for Fund benefits, but may be covered under the Workers' Compensation program.

Costs related to medical care, hospitalization, prescriptions, therapy, and transportation will be paid when an occupational injury or illness is "directly connected with operations as a commercial fisherman" in Alaska waters or on shore preparing or dismantling boats or gear used in commercial fishing. In addition to the expected cuts, sprains, and fractures, other examples of covered conditions are hernias, arthritis, and traumatic sciatica. Bronchitis, pneumonia, and pleurisy caused by or aggravated by the fishing activity are also included.

By statute, certain conditions are generally not covered—for example, the common cold, flu, cancer, appendicitis, or smoking-related conditions. Quite specifically, recreational drug or alcohol-related injuries are excluded from coverage. Conditions caused by not following good hygiene and health practices or improper care are not covered.

The Fishermen's Fund report is considered the fisherman's claim form or application for Fund benefits. When signed and dated, it is considered an affidavit attesting to the validity of the claim. When properly filled out and submitted with requested materials, such as a copy of the crew member's license, Fund staff can very quickly process the claim and reimburse the fisherman or pay the care provider. In FY 2003, 696 claims were submitted (Table 5). Approximately two-thirds (389) of all approved claims (583) were approved the same day as received by office staff.

In general, the most common delay in the approval process is waiting for an explanation of benefits from an insurance company to document why a claim is not covered by other insurance policies. This quite often causes approval delays of 150 days. Recall that the Fund is considered an emergency fund that is the payer of last resort. When the administrator cannot approve an

Table 5 : Outcome of claims

Approved claims	583
Claims approved by council	7
Pending claims	16
Claims pending council	25
Claims paid by primary insurance	14
Claims to payer of last resort	12
Withdrawn claims	3
Claims denied by council	33
Miscellaneous	3
Total	696

application for benefits, it must go before the Advisory and Appeals Council for review, then denial or approval. Common reasons for delays that require council review include no evidence of a license at the time of injury or illness, no response to inquiries from the staff about items on the application, or not related or connected to the operations of a commercial fisherman in Alaska.

Health care providers can be very helpful in the paperwork process and often file on behalf of the individual after obtaining proper information and signatures. Since they are the ones usually owed for services, it is to their advantage to understand the system. Ultimately, it is the injured party who is receiving treatment who is responsible for payment. An awareness program by the Fund administrator has sought to make the claims process more easily understood by the staff of various hospitals and clinics, Applications and information brochures are sent out annually to providers. Individuals can talk directly with staff via a toll-free phone line. A Web site (http://www.labor.state.ak.us/wc/ffund.htm) and e-mail (FishFund@labor.state.ak.us) are additional means of communication. At the present time, applications and documentation must be submitted as hard copy, but forms are available on-line.

Results

Table 6 shows the claims filed by gear type and fishing district for fiscal year 2003. These data are gathered from the application and point out several things. Nearly all of the salmon caught commercially in Alaska are taken with gill nets, seines, and trolling gear. Collectively, the salmon fisheries employ the largest number of persons in the commercial fishing fleet. As one

Table 6: Claims by gear type and district

Gear type	Fishing district					District unknown	No. of claims
	1	2	3	4	5		
Gill net	6	21	74	58	4	2	165
Seine	29	32	35	2	3	2	103
Troll	13	32	4	1	0	1	51
Pots	18	26	37	7	5	0	93
Trawl	2	6	18	1	0	0	27
Longline	15	34	85	13	4	4	155
Not stated	24	15	36	14	6	5	102*
Total							696
* Two were not from Alaska							

would suspect, they account for the largest number of claims to the Fund, and these most often occur in the summer months. With a longer season, longliners have more exposure and show claims in their most significant areas of operation. District 5, including areas in western and northwest Alaska, has relatively few commercial fisheries and participants and thus shows very few claims. Subsistence use of fishery resources is very high in this area, but this activity is not covered by the Fund.

From the statistical point of view, Fishermen's Fund injury data are far from comprehensive and complete. First, not everyone who is injured during commercial fishing operations files a claim with the Fund. Some know their own personal health insurance will cover the cost. Others have coverage provided by the Indian Health Service, Veterans' Affairs, or the federal Medicare program. Some individuals and their health care providers are unaware of the program.

If a serious injury occurs that exceeds the standard $2,500 amount paid by the fund, the owner of a vessel usually files a claim directly with his vessel's Protection and Indemnity (P&I) insurance policy. The Fund does not pay the typical $2,500 or $5,000 deductible on these policies. On the other hand, the owner-operator of a vessel is typically excluded from a vessel's P&I policy and is likely to file with the Fund. Operators that choose not to have P&I insurance on their crew would be likely to encourage them to file with the Fund.

Data collection by the National Institute for Occupational Safety and Health

(NIOSH) at its Anchorage field office indicate that very serious accidents in Alaska are best documented through the Alaska Trauma Registry (Simonsen 1994). Significant admissions to emergency rooms and hospitals around the state are indicated on this registry and identify some of the occupational circumstances associated with the injury. For example, some of the serious crushing injuries that have occurred in the crab fisheries in the Bering Sea show up on this registry and never appear as Fishermen's Fund claims. The large and expensive distant water vessels are well insured, anticipating large medical bills, and potential lawsuits go directly to their insurance companies.

While the raw data covering injury and illness claims to the Fund cover a period of more than 50 years, they are not statistically complete to properly represent the spectrum of injuries and illnesses that befall those in Alaska's commercial fisheries. Especially in recent years, with the escalating costs of health care in the United States, there is more likelihood of minor events being heavily represented in the data to the exclusion of more significant events. Still, for anecdotal evidence of the nature of injuries and illnesses, the data can point to problems that often happen to those working on fishing boats. For example, lacerations caused by knives used to clean fish or mend nets are common. Slips and falls that occur on the same level, resulting in sprains or contusions would be expected on rolling, slippery decks. Again, because of the motion of the ocean and the heavy nature of much of the work on deck, overexertion injuries, strains, and sprains are common. Bodily reaction injuries, e.g., from bending repeatedly or slipping without falling, are common. Interestingly, according to the annual Workplace Safety Index compiled by Liberty Mutual (North Pacific Fishing Vessel Owner's Association 2003), these injury causes accounted for 51% of workers' compensation direct costs in 2000 for all workplace injuries nationwide. So perhaps while not as dramatic as an amputation, the cumulative numbers represented by these types of injuries become very expensive in the long run. In the short term they are certainly disruptive to the work on a vessel, often causing an unwanted trip to port and a loss of productivity. Needless to say, the problems for the individual injured should never be minimized.

To illustrate how the Fishermen's Fund helps individuals, let me present a typical example. While pulling in gill net gear, crewman Bob Brown's finger gets tangled and dislocated. The captain pulls into port, and Bob goes to the emergency room to get the finger x-rayed. The doctor says the finger is not broken, just dislocated, tapes it up, and sends Bob back to the boat.

The bill is $300 for the emergency room visit, $150 for the x-ray, and $200 for the doctor's fees, for a total of $650. Under the Jones Act, the captain, John Smith, is responsible for the payment of the injury. He informs the hospital that the Fishermen's Fund will pay for the injury. The hospital has Bob Brown fill out the Fishermen's Fund claim form, and the physician fills out his portion. A copy of Bob's crew license, the medical bills, and chart notes are attached and mailed to Fishermen's Fund.

Fishermen's Fund receives the paperwork. That day they validate the license and review the claim for missing information and eligibility. If everything is there, the claim is approved for payment. Information is entered into the computer, and a payment request is sent to the fiscal department. Bob Brown's medical bills are paid, the captain is off the hook for payment, and P&I insurance is not held liable.

At times the care and treatment of a person's injury or disability extends beyond 1 year from the date of initial allowance. Furthermore, in today's world of high medical costs, bills may quickly exceed the $2500 originally specified in the 1951 law creating the Fund. Individuals have the right to request in writing an extension of the time and/or the benefit limits. They must give compelling reasons for justifying the request and cite the "amount of relief" and "the extent of additional time" required (Fishermen's Fund). The Council reviews information submitted by the fisherman on his or her financial status. It also considers the impact of the injury or illness on the fisherman's ability to earn a living while undergoing treatment and to continue to earn a living as a commercial fisherman. Each case is carefully considered by the Council, and a decision is delivered to the individual. All decisions of the Council may be appealed to the Commissioner of Labor and Work Force Development if the claimant feels he or she has not been treated fairly.

The following examples will illustrate the benefits of this provision.

(1) On the last set of the day, the captain is helping the crew pull in seine gear. His arm gets entangled in the deck winch and is severely dislocated. The crew immediately heads to port and takes the captain to the hospital. The doctor looks over the captain and suggests shoulder surgery in the next few days.

The captain is off work for 2 weeks while he is recovering from surgery and beginning physical therapy. His bills are running somewhere around

$15,000, and he has not been able to finish the fishing season. His P&I insurance only covers the crew and not him. He has no personal medical insurance.

He requests that the hospital submit the claim to the Fishermen's Fund. Fishermen's Fund receives the claim and verifies the valid permit and that the injury was directly related to commercial fishing. Fishermen's Fund pays to the $2,500 limit and mails a letter to the vessel owner asking if he would like to apply for an extension of benefits.

The vessel owner fills out additional paperwork and submits it to Fishermen's Fund. Fishermen's Fund Advisory and Appeals Council reviews the claim, deliberates on the case, and approves the $12,500 extension. Medical bills are paid, the vessel owner is healed, and he is not in debt with the hospital.

(2) One woman, also an owner-operator, suffered an umbilical hernia while working on deck. Though she had private health insurance, she still had to pay a significant deductible, and many of her surgical and medical expenses were only paid at 80%. In addition to the typical $2,500, she received an additional $680 toward expenses that were her responsibility as described in the explanation of benefits.

Discussion

The Fishermen's Fund has not kept pace with inflation. Whereas in the 1950s, a benefit of $2,500 was capable of providing significant assistance to sick or injured fishermen, nowadays it rarely covers more than a doctor's visit, basic emergency room treatments, or basic therapies. It may be that individuals are better covered by vessel P&I coverage and private health insurance than in the past. There are still numerous individuals who fall through the cracks and cannot cover their expenses when an injury or illness occurs during commercial fishing operations. These individuals and the care providers who are owed for their services are the ones who benefit to the largest extent from the Fund.

It would be tempting to ask the State of Alaska Legislature to allow a higher limit than $2,500 to be paid for injuries or disabilities. Recall that this Fund predates statehood and is one of the few dedicated funds allowed in the state government. Nearly all state appropriations are currently made through the state's General Fund and are subject to the wishes of the legislature and the potential veto of the governor. Given the financial difficulties that the state is currently wrestling with, any change in the original intent of or language establishing the Fund could easily mean the loss of its grandfather status and its incorporation into the General Fund. It is conceivable that the Fund would be financed at considerably lower levels or be eliminated altogether, with the revenues used for other purposes. The losers would be the more than 700 individuals who annually receive financial benefits and providers, often in remote locations, who might have difficulty collecting payments from financially strapped individuals.

Conclusion

The Alaska Commercial Fishermen's Fund is a small but significant program with a 50-year history of service to the fishing industry. It is a fine example of an effort to ensure that by virtue of their work, fishermen have access to medical care. Currently, it is self-sustaining and well-managed. Future efforts at streamlining service with updated computer programs and the potential for on-line filing will be helpful to some fishermen and many providers. Continued communication with providers and members of the industry will ensure that the program is well utilized by commercial fishermen working in Alaska.

References

State of Alaska. Alaska Adminstrative Code, Chapter 55: Fishermen's Fund. 8 AAC 055.010-040.

State of Alaska. Alaska Adminstrative Code, Title 23: Labor and Workers' Compensation. AS 23.35.010-150.

North Pacific Fishing Vessel Owners Association (2003). Fishermen's Fund: Requesting More Benefits. *Vessel Safety Program Newsletter* 5.

Simonsen B (1994). Alaska Trauma Registry. *In* Proceedings of the National Fishing Industry Safety and Health Workshop (Anchorage, Alaska, Oct. 9-11, 1992), Myers ML and Klatt ML, eds. Cincinnati, OH: National Institute for Occupational Safety and Health. DHHS (NIOSH) Pub. No. 94-109.

THE FAROE ISLANDS AND FISHING

Óli Jacobsen
Chairman of the Fishermen's Union
Faroe Islands
e-mail: oli-fisk@post.olivant.fo

First of all, I will say how happy I am to be here: I attended the first International Fishing Industry Safety and Health conference three years ago and found it very useful. Therefore, I wanted to come back, and this wonderful spot, Sitka, which I never have heard of before, is an ideal choice of site for this conference. This time I am not alone. We found the first conference so interesting that this time I brought three more people from my country. The Faroese delegation is one of the largest international groups here at IFISH II, outside those people coming from North America. The amount of time and travel expenses for our delegation to get to Sitka, however, can be justified given the importance of fishing in our country.

The Faroe Islands consist of 18 islands, located between Iceland and Scotland. The territory is 1400 sq km, with a total population of 48,000. The islands are an autonomous part of the Kingdom of Denmark, with our own culture, parliament, and government. We have our own language, but Danish and English are our second and third languages. All the younger people today speak English.

I became a fisherman in 1958, at the age of 14. Since 1971, I have been the elected chairman for the Faroese Fishermen's Association. I can say that I have had a practical and organizational association with the Faroese fishing industry over the last 45 years, and I have experienced all the great changes in the fishery during this time. Most notably, these include improvements in the area of safety. Members of our union have some of the best safety records in the industry, which makes us very proud.

Our union

The Faroese Fishermen's Association was founded in 1911, and currently has about 2,500 members who work on the high seas fleet. Our purpose is, of course, to work on behalf of the interests of fishermen and to promote safety and health as high priorities.

Our fishermen are paid according to collective agreements between the ship owners' association and our union. Faroese fishermen are paid with a share of the catch. In addition, our state guarantees fishermen minimum wages equal to the daily pay of an unskilled laborer who works 8 hours a day.

Our union has a newspaper, the Ffbladid, which is issued every other week. Our paper has a very high reputation and is widely read throughout the Faroe Islands. We write regularly about safety at sea, and as the paper is sent on board all fishing boats, the message finds its way to all fishermen. We plan on publishing a complete report on this conference, which will also be shared with our readers.

Fishing–the main industry of the Faroe Islands

The main industry in our country is fishing; there are about 3,000 commercial fishermen in the islands out of a total population of approximately 47,000. Fish represents nearly all goods exported from the Faroe Islands. Few countries are as dependent on the sea's resources as ours. If there is a crisis in the fishing industry, the entire Faroe Islands community suffers.

The Faroese fishing industry began 130 years ago with the adoption of decked fishing boats. Before that, agriculture and fishing from open boats had characterized the Faroese economy. Just prior to the Second World War, the Faroes had 170 fishing vessels amounting to 19,000 GRT. During the war, the 39 ships and 180 men who were lost badly affected the country–not only in human terms, but also economically. In the 1950s, we started to build up a new fishing fleet. Since the 1960s, the Faroese fishing fleet has been among the best-developed in the world.

Traditionally, the Faroese fishing industry has been based on distant-water fishing, far from our shores. Our ships were, from the very beginning, and are fishing all over the North Atlantic–from Canada in the west to the Barents Sea in the east and north. In the past, the catches were salted and later frozen on board, and the crew stayed away from home for months, as there was no catch limit, but now the trips are limited to 3 months.

Our fishermen are not only fishing on Faroese boats; they have been crewing fishing fleets of other countries for decades. In the 1960s, for example, a great plan was promoted in Greenland to build up an ocean-going trawler fleet for the North Atlantic cod fisheries. Later those trawlers were converted to catch and freeze prawns. Faroese were involved in the planning and execution throughout these projects, and many Faroese now sail as skippers and engineers in this fleet. A similar situation occurred, with Faroese assisting and crewing, when the Canadians began building up their fleet of prawn trawlers. (Faroese are working all over the world, and not only in fishing. Tomorrow I will visit a cousin of my father on the other side of the border, in Canada. He had a leading job in an aluminum plant for many years.)

After the many industry changes in the 1970s, commercial fishing took place primarily within our domestic waters, where the catches are iced on board and delivered ashore every 1 or 2 weeks. But the Faroese still conduct commercial fishing all over the North Atlantic Ocean.

Faroese commercial fishing management system

We think we have the best fish stock management system possible. Instead of quotas, we have effort limitation. We tried a quota system in the early 1990s, but we found this system didn't work in mixed fisheries. We also realized biologists' predictions regarding the marine resource were not always accurate, which also impacted the efficiency of a quota system. As a result, in co-operation with biologists, fishery administration, and the industry, we created a effort-based system.

The ships are allocated a number of days at sea in which to conduct fishing. Under this allotment scheme, they are more likely to bring ashore the entire catch, and waste and discard are less of a problem than under other systems. In addition to limitations on days at sea, we have closed areas and have incorporated other technical limitations.

The system appears to be working. Since it was implemented in 1996, we have had our most stable period of fishing ever. We still must address disagreements with our biologists, who have not officially recognized this Faroese system as a viable management system for fisheries. Marine biologists continue to use the quota system in their estimates, and independently of the state of the fish stocks, they recommend every year that efforts be cut by nearly one-third. However, the commercial fishing industry continues to adhere to the effort limitation method of managing our fisheries, and in so doing, has begun to demonstrate that fishermen share a common interest with biologists in maintaining sustainable fisheries. There is increasing interest from other countries in our management system. Representatives from the fishing industries in other countries are coming to the Faroes to learn more about our system, recognizing that the traditional quota system is not necessarily the only solution for fisheries management.

Other fishing-related industries in the Faroes

We have quite a large industrial base related to commercial fishing on the islands. Processing of fish fillets for freezing began in the 1950s, but this industry did not develop much until 1977, when the Faroese fishing limit was extended to 200 miles. There are now about 100,000 tons of ground fish landed at those factories yearly, providing jobs to shore-based workers, especially women.

I want to mention our fish farming, especially of salmon, which we started up in the 1980s. After rapid growth of this sector, there have been serious setbacks during the last couple of years because of the fall in the prices of farmed salmon, as well as an increase in disease in farmed stock. We hope the industry will recover after a couple of years, as the farming is the only real alternative we have to fishing on the ocean. In addition to the ground fish fleet, we have also a pelagic fleet with a number of big purse seiners, profitable both to fishermen and ship owners.

Safety and health of fishermen

The Faroese Fishermen's Association wants to improve safety conditions for our fishermen. Safety has been improved compared to the old days, when ships just disappeared with the entire crew–typically resulting in the deaths of 20 fishermen coming from the same village, often from the same family. We recognize that many improvements have been implemented, but we also recognize that more education and training in this area must occur. It is very important to address the human factors that are associated with injuries and fatalities to commercial fishermen. In the Faroe Islands, it is compulsory for fishermen to attend a safety course before they go fishing, and new boats must be equipped with safety-related devices and gear. With reference to the discussion about deck safety earlier during the conference, I would like to add that nearly all sea-going vessels are over-decked, which results in a considerable improvement in working conditions.

There is still need for further improvements—first, with regard to safety, but also concerning the living and working conditions on board. I am of the opinion that more attention should be paid to questions regarding lifestyle factors on board, as discussed at this conference by Anna Maria Simonsen. I became acutely aware of that issue when I was brought to the cardiovascular unit at the hospital 5 years ago. Many of my fellow patients were fishermen or former fishermen. The chief doctor confirmed that this pattern for hospitalization of fishermen had been occurring for quite some time. According to the cardiologist, fishermen are clearly overrepresented in heart diseases compared to the general Faroese population. They also suffer from hypertension, had an increased body mass index, are heavy smokers, and have unhealthy diets. We want to bring your attention to this question. We have taken steps to more fully define the lifestyle and health issues affecting our commercial fishermen, and funds have been given for further studies of the question. Our union has decided to provide the crew of each ship a new book with prescriptions for healthy food. We have also, in our collective agreement with the ship owners union, agreed to a nonsmoking policy aboard commercial fishing vessels.

We look forward to bringing home the experiences from this IFISH II conference. We have found the program very interesting, and we are eager to bring home all the input we have got. I am happy for my stay in Sitka. I have met many people. I have spoken to fishermen from all around the world and I am sure that the Faroese fishermen will benefit from what we have learned here.

COMMERCIAL FISHING SAFETY: MAGNITUDE OF PROBLEM, RISK FACTORS, AND POTENTIAL SOLUTIONS

Jennifer M. Lincoln, MS
National Institute for Occupational Safety and Health
Alaska Field Station
Anchorage, Alaska, USA

Background

Commercial fishing is one of the most dangerous occupations in the United States. In 2002, commercial fishermen had the second highest traumatic injury fatality rate of all workers—71.1/100,000 workers, which is 16 times the national rate of 4.4/100,000 workers across all occupations (Bureau of Labor Statistics 2003). Only timber cutters had a higher fatality rate of 117.8/100,000 (Bureau of Labor Statistics 2003). Many fishermen work in isolated locations and harsh environments with high winds, cold water, icing conditions, and long work days. They suffer fatigue, physical stress, and financial pressures to push their vessels and crew to make their living (Lincoln and Conway 1997; Conway, Lincoln et al. 2002).

Since 1991, many activities have been implemented to monitor and improve the safety of this industry. The purpose of this paper is to discuss the magnitude of the safety problem and to discuss some particular interventions that could be exported to other parts of the country and to other fishing countries to improve safety. This paper will discuss the problems of vessel sinkings, deck injuries, and falls overboard, and potential interventions for each of these problems.

Vessel sinking

Most fatalities that occur in the commercial fishing industry in the United States are due to the loss of a vessel. From 1994-2000, 907 vessels sank in the United States, resulting in 218 fatalities (Dickey 2003), an average of 130 vessels and 31 lives lost each year.

Accurate workforce estimates are not available for the country to use to calculate trends. However, in Alaska, where such estimates have been made, it has been shown that there has been a significant decline in commercial fishing fatalities. This decline has occurred primarily in events related to vessel sinkings. From 1991-1999, an average of 34 vessels were lost each year in Alaskan waters with 106 people on board these vessels. The case-fatality rate has decreased from an average of 22% in 1991-1993 to 6% in 1997-1999 (Conway, Lincoln et al. 2002).

The US Coast Guard has developed several programs to prevent fatalities due to fishing vessel sinkings. The three that will be discussed include damage control trainers, stability trainers, and dockside enforcement activities.

Damage control trainers

The damage control trainer is used to simulate flooding situations on fishing vessels. Fishermen practice controlling flooding using plugs, rubber, and other miscellaneous items that would be found on a fishing vessel or in a damage control kit on the vessel. Being able to control these flooding situations could allow fishermen to save the vessel long enough to get to port or await aid. These damage control trainers have been used across the United States to train thousands of fishermen (Society of Naval Architects and Marine Engineers 2003).

Stability trainers

Stability trainers are used by the Coast Guard to educate fishermen on the effects of operational decisions on vessel stability. These vessel models are built to 1/16th scale of an existing fishing vessel. There are four cargo holds, a lazarette, and an engine room that can be filled with water, allowing the fisherman to observe adverse effects on vessel stability. The trainer can also be used to show how an increase in the center of gravity affects stability (Kvaerner Masa Marine 2003). These trainers enable the complicated dynamic of stability to be illustrated with a vessel model.

Dockside enforcement

Dockside enforcement strategies have been implemented for specific fleets in the Pacific Northwest. These efforts have been shown to be effective in improving the safety of specific fleets such as the Bering Sea crab fleet (Medlicott 2002) and the Oregon crab fleet (Lawrenson, Farrell et al. 2003). Fishing vessel examiners in Alaska, Washington, and Oregon have developed targeted operations called "pulse operations." In Alaska, they have boarded Bering Sea crab vessels prior to the crab season opening to review the vessel's stability booklets and make sure they are not overloaded with too many crab pots (700-pound cages that are used to catch crab on the ocean floor). The examiners make sure boats are loaded properly for the predicted weather conditions and look at the safety gear, including the life raft, electronic position-indicating radio beacon (EPIRB), and immersion suits, to ensure there are enough for the crew and that everything is installed properly.

If the examiners find that the pots are loaded incorrectly or if there is a problem with the safety gear, then the vessel may not be allowed to get underway until the discrepancy is fixed (Medlicott 2003). Since implementation in 1998, only one fatality has occurred in this fishery, which was due to a fall overboard. The crab industry strongly supports this initiative (Medlicott 2002).

Washington and Oregon examiners conduct similar operations on the Dungeness crab fleet along their coast. These are smaller vessels that many times do not have stability letters, so in this operation, the examiners just look at safety equipment (see Lawrenson , Farrell, and Hardin, this proceedings). Safety gear suppliers have reported that fishing vessel owners and operators are reacting to this annual operation by getting gear ordered and checked earlier. The Coast Guard believes there has been a change in fishermen's behavior because of these inspections (Lawrenson 2003).

In addition to these Coast Guard projects, training programs are available where fishermen get appropriate emergency training on how to react to emergencies at sea. These courses cover several topics including MAYDAY calls, EPIRBs, immersion suits/personal flotation devices (PDFs), life rafts, flares, emergency drills, and firefighting. The North Pacific Fishing Vessel Owners Association, based in Seattle, Washington, has classes on safety equipment and survival procedures, emergency drill instruction, fire prevention, and vessel stability and damage control (North Pacific Fishing Vessel Owners Association 2003). The Alaska Marine Safety Education Association (AMSEA), with instructors around Alaska, also offers emergency drill instructor courses. A study evaluating the effectiveness of AMSEA's marine safety training showed that these courses were effective in preventing drownings among Alaskan commercial fishermen (Perkins 1995).

Deck injuries

Not all fatalities are due to vessel sinkings. The fishing vessel is often a congested work area with hydraulic machines and fishing gear. About 10% of the fatalities in this industry nationwide are due to these types of deck hazards (Dickey 2003). Nonfatal injuries are also primarily due to deck injuries. Surveillance for nonfatal injuries in the commercial fishing industry is problematic. Although there is a requirement that severe injuries (loss of work for 3 or more days) be reported to the Coast Guard, they do not investigate or necessarily keep track of nonfatal injuries among fishermen. In Alaska, however, the state's Department of Health and Social Services maintains the Alaska Trauma Registry. This registry contains information on all hospitalized injuries, including those that occurred in the commercial fishing industry. A paper by B. Husberg, J. Lincoln, and G. Conway in these proceedings gives a thorough description of these data.

Based on these findings, the Deck Safety Project was established to examine the relationship between the vessel, fishing equipment, and the fishermen. We have many partners on this project, including Jensen Maritime Consultants. Many interventions have been identified to reduce the risk of injury at sea. These are discussed more in E. Blumhagen's paper in these proceedings. The solutions highlighted include ways of controlling fishing gear, identifying hazards on deck, and visibility. The Deck Safety Project is continuing to study the causes of these deck injuries and appropriate strategies to prevent them.

Falls overboard

Falls overboard are caused by being washed overboard by waves, slips, trips and falls on deck, or being pulled over by fishing gear. From 1994-2000, 135 fishermen were killed due to falls overboard in the United States (Dickey 2003). Falls overboard accounted for 29% of all commercial fishing fatalities. To prepare this description of the falls overboard problem in the US commercial fishing fleet, I reviewed these 135 cases. Events were categorized by fishing gear, geographic location, number of crew on board, length of vessel, and fishery.

Fatalities due to falls overboard occurred most often along the Gulf Coast (49, 36%), followed by Alaska (24, 18%), and New England (23, 17%). The highest number of falls overboard fatalities occurred using towed or dragged gear (55, 41%), followed by static gear (34, 25%). The most common fisheries involved were the shellfish fisheries: shrimp (42, 31%), crab (23, 17%), and lobster (17, 13%). In 36 of the fatalities (28%), the victims were fishing alone.

Several factors associated with preventing deaths from falls overboard have been identified. Strategies to prevent fatalities related to falls overboard include (1) avoiding becoming entangled in fishing gear by line lockers, bins, and fairleads, (2) interrupting the force of being pulled over by cutting the engine or cutting the line, (3) re-entry into the vessel after entering the water, and (4) use of personal flotation devices to aid persons in the water (Backus et al. 2002). Other interventions include shelter decks or seawalls to protect fishermen from weather, barriers between fishermen and gear to prevent entanglements, and practice with rescue gear for quick retrieval of victims in the water. Clearly, further investigation is needed of fatal events, as well as additional studies that help identify protective factors for fishermen who were successfully rescued from falls overboard.

The Seventeenth Coast Guard District in Alaska lists "wearing PFDs at all time while on deck" as one of their "Ready for Sea" checklist items that all fishermen should meet before going out to sea. "The practice of wearing PFDs while working on deck … is a standard-of-care vessel crews should adopt" (p. 2002). A survey by the Coast Guard and NIOSH's Alaska Field Station showed that 88% of the skippers on crab boats require their crew to wear PFDs when climbing on the stack (stack of crab pots on the back of the deck), but only 13% of them require wearing PFDs while working the gear (Thomas, Lincoln et al. 2001). A review of all drowning incidents subsequent to commercial fishing vessel losses showed the effectiveness of PFDs and survival suits. The study found that survivors of events in which at least one person drowned were 8.3 times more likely to have been wearing a PFD or immersion suit than were those who died (95% CI 3.59-19.24) (Conway, Lincoln et al. 2002).

Summary

Several projects and ideas have been identified as ways to improve safety in the commercial fishing fleet. Interpreting surveillance data to help develop such programs is important. Collaborations have proven to be effective in developing ways to increase safety in the industry. Fishery-specific approaches such as the Dockside Enforcement Project and the Deck Safety Project can be applied to other areas. It is important to identify more programs, tools, and training programs to continue progress in making this industry safer.

References

Backus AS, Brochu PJ et al. (2002). Work practices, entanglement of lobstermen, and entanglement prevention devices in the Maine lobster fishery: A preliminary survey. *In* Proceedings of the International Fishing Industry Safety and Health Conference (Woods Hole, Massachusetts, Oct. 23-25, 2000), Lincoln JM, Hudson DS et al., eds. Cincinnati, OH: National Institute for Occupational Safety and Health. DHHS (NIOSH) Pub. No. 2002-147.

Bureau of Labor Statistics (2003). Washington DC: U.S. Department of Labor

Conway GA, Lincoln JM, Hudson DS, et al. (2002). Surveillance and prevention of occupational injuries in Alaska: A decade of progress, 1990-1999. Cincinnati, OH: National Institute for Occupational Safety and Health. DHHS (NIOSH) Pub. No. 2002-115.

Dickey D (2003). Analysis of fishing vessel casualties: A review of lost fishing vessels and crew fatalities, 1994-2000. US Coast Guard.

Kvaerner Masa Marine (2003). KMM Web site.

Lawrenson K (2003). Personal communication, Portland, Oregon.

Lincoln JM and Conway GA (1997). Commercial fishing fatalities in Alaska: Risk factors and prevention strategies. Cincinnati, OH: National Institute for Occupational Safety and Health. Current Intelligence Bulletin 58. DHHS (NIOSH) Pub. No. 97-163.

Medlicott C (2002). Using dockside enforcement to compel compliance and improve safety. *In* Proceedings of the International Fishing Industry Safety and Health Conference (Woods Hole, Massachusetts, Oct. 23-25, 2000), Lincoln JM, Hudson DS et al., eds. Cincinnati, OH: National Institute for Occupational Safety and Health. DHHS (NIOSH) Pub. No. 2002-147.

North Pacific Fishing Vessel Owners Association (2003). Web site.

Page EE (2002). Ready for sea: The Seventeenth Coast Guard District's safety program. *In* Proceedings of the International Fishing Industry Safety and Health Conference (Woods Hole, Massachusetts, Oct. 23-25, 2000), Lincoln JM, Hudson DS et al., eds. Cincinnati, OH: National Institute for Occupational Safety and Health. DHHS (NIOSH) Pub. No. 2002-147.

Perkins R (1995). Evaluation of an Alaskan marine safety training program. *Public Health Reports* 110 (6): 701-702.

Society of Naval Architects and Marine Engineers (2003). Web site.

Thomas TK, Lincoln JM, et al. (2001). Is it safe on deck? Fatal and nonfatal workplace injuries among Alaskan commercial fishermen. *American Journal of Industrial Medicine* 40: 693-702.

Session Twelve

SESSION TWELVE:
MIXED CATCH

Small vessels and big waves in the Bay of Bengal (Photo courtesy of Yugraj Yadava)

Session Twelve

TAMING *MAL DE MER*: A REVIEW OF CURRENT KNOWLEDGE ON THE PREVENTION AND TREATMENT OF SEASICKNESS

Terry Johnson, Marine Advisory Program Agent
University of Alaska Fairbanks Sea Grant
3734 Ben Walters Lane 205
Homer, Alaska, USA
E-mail: rftlj@uaf.edu

Introduction

The English words *nautical* and *nausea* both derive from the old Greek word for "ship," and a close association between the two continues to this day.

Seasickness is a debilitating condition that can afflict almost anyone, sometimes even in relatively mild sea conditions. Its direct effects are temporary and nonfatal, but it may lead to potentially dangerous dehydration and can create a hazardous situation by robbing crew members of the ability—and sometimes the will—to care for self and ship. Some individuals appear to be relatively immune while others, including many commercial fishermen and professional mariners, suffer from it intensely and repeatedly. Aversion to seasickness is a major deterrent to recreational boating, sport fishing, and cruise vacationing.

Although fear may be a contributing factor, the condition itself is neither psychological nor "all in the head." It is a very real physiological response to disturbance of the normal sense of balance, as perceived via the visual, vestibular (inner ear), proprioceptor (peripheral body sensory), and somatosensory (overall body sensory) systems.

Seasickness is largely preventable and is treatable once in progress. Many people with motion sickness susceptibility function comfortably at sea by using good information to apply appropriate behavior and effective medication.

Purpose

The maritime popular press and dock-talk frequently address seasickness prevention and treatment, and both present anecdotal evidence to support the effectiveness of various pharmaceuticals, devices, and folk remedies. Some discussions are enlightening, others downright silly, but few are based on empirically derived information. At the same time, government agencies, including the National Aeronautics and Space Administration (NASA), universities, research hospitals, and navies have conducted controlled experiments pertinent to motion sickness suppression, but their results have not been widely disseminated. Many marine, water sports, and medical writers have offered seasickness advice based on lengthy experience on the water. Some of that advice is confirmed by the research results, while much appears intuitively valid but not clinically tested.

Individual mariners, vessel owners, and passengers need the best available information on controlling seasickness. The purpose of this study is to review the current state of knowledge, help the industry sort out valid from unsubstantiated information, and suggest approaches that may help mariners minimize the effects of this malady.

Methods

The author reviewed 179 published sources on seasickness prevention, control, and treatment. The categories of sources are as follows: medical or scientific journal articles (15); medical journal article abstracts (23); drug information sheets from the American Society of Health-Systems Pharmacists (ASHSP) and other on-line pharmacy sites (12); drug manufacturer information sheets (7); alternative remedy information sheets (9); Web postings from physicians, medical schools, consultants, drug companies, and letters from physicians (20); comments gleaned from chat room postings (40); articles in newspapers, popular magazines, maritime recreation and trade publications, and marine recreation and industry-related Web sites (53). Additional input was obtained through personal communications with a number of researchers, drug company representatives, physicians, and individual seasickness sufferers.

Journal articles and abstracts reported on research conducted in the United States, Canada, United Kingdom, Switzerland, the Netherlands, Germany,

Denmark, and Israel. Nearly all of the other materials came from sources in the United States.

Medical and scientific journal articles and abstracts are believed to be based on legitimate, peer-reviewed research. Drug manufacturer information sheets and drug information sheets from ASHSP are believed to be factual descriptions of the contents, effects, and side effects of drugs. Postings by physicians, medical schools, and consultants are believed to be legitimate analyses by qualified medical professionals, but are not peer-reviewed nor research-based. Comments from chat rooms are just that—accounts based on personal anecdotal experience. Articles in newspapers, magazines, and on industry-related Web sites run the gamut from well-researched journalistic endeavors, to humor pieces, to thoughtful essays, all summarizing decades of personal experience.

Sources addressed causes of seasickness, behavioral prevention and treatment, pharmaceuticals for prevention and treatment, alternative therapies, and comparisons among competing drugs and devices. No attempt was made to ascertain or score the legitimacy of the sources, nor was there any sort of statistical analysis of the messages they contain. Rather, the author endeavored to find verification of results through multiple references and corroboration between science-based and experience-based sources.

The author also attempted to correlate results obtained through this review with the experiences of passengers aboard his own wildlife charter boat, which operates on the frequently tempestuous waters on Bristol Bay, Alaska, during the summer. The number of clients observed annually is too small to create a statistically significant sample, but casual observation has been sufficient to cast doubt on some published claims and to corroborate others.

What follows is a synthesis of the current state of knowledge on the control of seasickness.

Results

What it is
Seasickness is a physiological condition characterized by dizziness, lethargy, nausea, vomiting, and a host of other unpleasant and often debilitating symptoms. It can result from a variety of sensory inputs and manifests itself

differently in different people. It is similar to other forms of motion sickness, and it strikes somewhat randomly. Some people are affected only in violent storm seas, while others succumb in the gentle roll of a ground swell.

It normally results from repeated rhythmic motion at the frequency range of sea waves. It is now clear that seasickness usually occurs when the vestibular and proprioceptor systems detect motion out of harmony with normal visceral activity. The emetic center of the brain interprets this input as evidence of poisoning and triggers the stomach's normal defensive action—to purge. In other words, seasickness is an inappropriate physiological response to certain physical stimuli (Money, James et al. 1996).

Much research on motion sickness has focused on the role of the vestibular system (inner ear), which plays a role in maintaining body orientation and balance. Persons with disease-damaged vestibular systems and animals whose inner ears have been surgically removed are virtually immune to motion sickness. They also don't exhibit normal nausea response to poisoning. Many of the anti-emetic (nausea preventing) drugs work by desensitizing the vestibular system.

Many sources state that seasickness is the result of dissonance between the vestibular and vision systems that often occurs when persons go below deck or focus their vision on objects inside the cabin. However, if it were caused simply by the difference between what the eye sees and the inner ear experiences, it could be prevented or cured simply by keeping the sufferer's vision focused on the sea outside the boat. Usually this is not the case.

Seasickness in most people has two components. One is distress—the sensation of malaise, dizziness, and nausea—that results from sensory inputs to the brain. The other is the stomach's reaction to it, which can be churning, acid overload, and vomiting. People may experience one without the other, but distress in the motion center of the brain seems to trigger a reaction in the stomach. Antacids and stomach folk remedies may ease the unpleasant gastric effects of seasickness, but will not attack the cause.

While the genesis of most seasickness is repeated rhythmic motion, especially up and down motion), it is exacerbated by odors, lack of ventilation, fear or apprehension, a sense of lack of control, and substances ingested by the sufferer, including alcohol, acidic drinks such as coffee and cola, and spicy or fatty foods. Even the sight of a vomiting shipmate can trigger nausea.

How to predict it

Nearly all people can become seasick in one situation or another. Nearly everyone gets over it ("gets their sea legs") after a period of anywhere from a few hours to a few days if they remain at sea that long. Some career mariners get it the first day out on every voyage (see note a). The literature suggests that susceptibility varies with age: kids 5 to 12 are most susceptible and adults over 50 are slightly less susceptible than others. People with good aerobic fitness are slightly more susceptible than those who are out of shape (Cheung, Money et al. 1990). People who get migraine headaches (most of whom are female) are more susceptible. Females apparently experience seasickness more frequently than males, but while women are more likely to report feeling seasick, they often cope better with nausea, complain less, and are less likely than males to attempt to curtail a voyage due to seasickness (see note b).

Anyone who has experienced other forms of motion sickness is a candidate for seasickness, but many who have not still get sick on the water, even after previous seasickness-free voyages. A good self-test for seasickness susceptibility is to read while riding as a passenger in a car on a winding road and in stop-and-go traffic.

Behaviors to prevent seasickness

It is more effective to prevent seasickness than to treat it. Following are some behaviors that can combine to reduce the likelihood of an onset of seasickness:

+ The night before, eat and drink moderately and get plenty of sleep; take a first dose of prophylaxis (see "Drugs" below).
+ The morning of embarkation, eat a small breakfast, avoiding rich or fatty foods, dairy products, or high-protein foods; avoid coffee or drink a minimal amount; take second dose of preventive medicine.
+ On the boat, take a position with a good outside view of water and horizon; stay low and near the center of the boat, if possible; avoid smells of the head, galley, fuel, and exhaust; avoid reading or doing work that requires close-range eye focus; move around and get fresh air; avoid alcoholic beverages and tobacco smoke; if possible, remain standing, move about, and "ride the waves" in a manner such that the head remains relatively stationary as the body moves with the boat; avoid abrupt or rapid head movements; try to take slow, rhythmic deep breaths to decrease symptoms of motion sickness.

If seasickness starts, acknowledge the condition and act on it. Take additional medicine (again see below), maintain a view of the horizon, and avoid going below.

If vomiting feels imminent, don't fight it. Some people feel much better after vomiting, but be sure to vomit over the side or in a location that causes minimal distress to others aboard.

Treat seasick crew members or passengers as sick people: Protect them from injury or loss overboard during vomiting; help them avoid dehydration by offering water or light fluids; and offer nutrients and electrolytes with light snacks or moderate meals of bland foods. Children, in particular, should be made to take plenty of fluids to avoid dehydration.

Alternative, folk, or "natural" remedies

Many nonpharmaceutical substances or applications have been recommended, including honey, peppermint, Vitamin B6, citrus fruits, mango, saltine crackers, biofeedback, "artificial horizon" glasses (Tassier 2003), rubbing alcohol, herbal drinks and salves, a half-cup of seawater, immersing feet in ice water, and holding a can of cold beer behind the ear (Drouin 2001). Most probably are harmless, and some sufferers claim to experience a measure of relief. However, these remedies have not been proven in research, and relatively few people report success using any of them.

Ginger, in the form of candied root, ginger ale, ginger cookies, tea, or powdered root in capsules, has many advocates, although its effect mainly is to settle the stomach rather than to suppress the nausea center in the brain (Holtmann, Clark et al. 2002). Many anecdotal reports support its effectiveness. However, the available published studies on the effects of ginger on nausea have produced contradictory results, and most that focused on seasickness have not shown a statistical difference between ginger and a placebo (see note c).

TravelWell, a British product, uses music, tones, and pulses to quell the vestibular system. It is an audiotape played on a personal cassette player with headphones. It has earned testimonials from Australia and around the UK. However, the vestibular and cochlear (hearing organ) systems function separately so there is no physiological reason that music or tones would ease distress originating in the vestibular system (Cheung 2003).

Acupressure, applied to the precordial channel 6, or *nei kuan* position on the inside of the wrist, produces relief in some people from several kinds of nausea, including seasickness. Products such as Sea-Band, Travelband, Queaz Away Bands, and BioBand, which have small plastic buttons or studs embedded in an elastic wrist band so that they press on the *nei kuan* position, are sold for seasickness prevention. They are inexpensive and have no side effects, so they are commonly carried aboard vessels. Clinical proof of effectiveness is sparse (see note d), although nonclinical tests and some field experience suggests that they work for some people (see note e).

A variation on the acupressure band is the Relief Band, an electronic device similar in appearance to a wristwatch that contains a battery and two electrodes that rest on or near the *nei kuan*. The device sends a user-variable transcutaneous electric current into the wrist, which in some people apparently desensitizes the vestibular system. It should not be worn by individuals with cardiac pacemakers. The manufacturer claims the device can both prevent and alleviate seasickness and provides testimonials. Limited field experience indicates that it works for some users (see note f).

As noted, clinical proof is lacking for most of the nonpharmaceutical remedies. Scientists believe that a combination of the placebo effect and habituation to the environment contribute to the beneficial affects attributed those alternatives (Cheung 2003).

Drug effectiveness

Most people rely on any of several popular pharmaceutical products to prevent or suppress seasickness. Most are taken as pills, capsules, or chewable tablets; some as suppositories; syrups or injections; or as a gel or a transdermal "patch." Most have proven in both laboratory testing and field experience to be at least somewhat effective for most users. Comparative tests have produced contradictory results; that is, in some tests one drug performed better than another and in other tests did not perform as well.

Overall effectiveness of drugs is highly variable. One often quoted figure is that proper application of drugs reduces the incidence of seasickness by 50% and another is that occurrence among persons predisposed to seasickness is reduced from 80% to 20% (see note g). All induce side effects, some mild and some severe. The less potent drugs tend to produce drowsiness and dry mouth, while more powerful compounds may produce prostatic hypertrophy, confusion, muscle twitching, rash, headaches, and other symptoms that may include lack of ability to concentrate on tasks and loss of muscle control.

Over-the-counter drugs

Most over-the-counter (OTC) drugs sold specifically for motion sickness prevention are based on one of three chemical compounds. Dimenhydrinate (original Dramamine, TripTone, Gravol), cyclizine (Marezine), and meclizine (Bonine, Antivert, Dramamine II) are all antihistamines, a class of drugs commonly used to treat the symptoms of allergies, but also are central depressants. They work by blocking the histamine receptors in the brain's emetic or vomiting center. They are moderately effective and have minimal side effects, usually drowsiness and dry mouth. The old standard dimenhydrinate (Dramamine) generally is considered to cause more drowsiness than other antihistamines (AHFS 2001); hence, the popularity of "nondrowsy" formulations, such as Dramamine II (meclizine).

Diphenhydramine (such as Benadryl) is an antihistamine that is marketed as an allergy remedy and an OTC sleeping pill, but it has characteristics similar to those listed above and is sometimes taken for seasickness when nothing else is available.

The key to success with OTC antihistamines is to begin treatment at least 2 hours prior to boarding the vessel and, preferably, taking an initial dose the night before. It may be necessary to maintain the level of the drug in the body by taking additional pills during a long voyage, as effectiveness may last as little as 4 to 6 hours. (The manufacturer recommends taking no more than two meclizine tablets in 24 hours.) Small or drug-sensitive individuals have the option of taking only one tablet.

Because the drugs can cause drowsiness, users should avoid alcohol, which would increase the sedative effect. Most of the antihistamines have not been proven to harm human fetuses, but manufacturers advise pregnant women to consult with their doctors before use. Some have been shown to cause birth abnormalities when given in very high dosages to laboratory animals. Manufacturers also warn that they can cause undesirable side effects when combined with other medications and may not be recommended for patients with certain medical conditions.

Prescription drugs

Other drugs are used to treat seasickness, some with different pharmacological bases. Following is a list of some of the more familiar anti-emetic drugs. Most are available in the United States only by prescription, if at all.

Scopolamine, marketed as Transderm-Scop and commonly referred to as "the patch," has achieved wide popularity and is considered by some researchers to be the best single anti-motion sickness drug available. Numerous clinical and field tests have been detailed in published reports and most, but not all, find it more effective than OTC antihistamines at preventing (not curing) seasickness. Introduced into the bloodstream via a small adhesive patch worn behind the ear, it remains potent for 48 to72 hours. Since most people acclimate to sea motion with less than 3 days of continuous exposure, re-application normally is not required. However, effectiveness diminishes during that time. It is a belladonna alkaloid, and side effects include pupil dilation, blurred vision, and dry mucous membranes. In a few people, side effects are more severe and potentially dangerous, including hallucinations and temporary psychosis in some elderly patients. Users must wash their hands after handling the patch and before touching their eyes. After prolonged use, some people experience withdrawal symptoms (Gahlinger 1999). The patch is supposed to be applied at least 4 hours prior to exposure to sea conditions, but peak serum levels are reached after 24 hours so an earlier application may be preferable. Although not recommended by the maker, some drug-sensitive individuals cut the patch in two and wear only one half at a time.

Scopolamine also is sold as a gel to be rubbed into the skin and in tablet form as Scopace. The manufacturer claims research shows that the tablet is twice as effective as the patch. It also claims that in the digestive system the tablet dissolves in minutes, and each dose is effective up to 8 hours. A prescription is required to purchase any form of scopolamine.

Promethazine (Phenergan), an antihistamine-based prescription drug, is reported to be very effective at stopping vomiting and may be indicated for severe seasickness. It emerged as the best anti-space-motion-sickness drug in a study conducted by NASA, with 80% effectiveness (Putcha, Berens et al. 2002). It works for both prevention and treatment, is available as a syrup, suppository, or injection (options that can make it useful to people who are too sick to take a pill) and is recommended for children as young as 2 years old.

Cinnarizine (Stugeron, Antimet) may be the most popular remedy in Europe and is considered more effective than OTC antihistamines while causing less sedation. It has proven effective in clinical trials. It is not FDA-approved for sale in the United States. Some users have expressed concern over its safety (Bohl and Anonymous 1999), and potential users should be aware that it comes in several dose levels that may produce different side effects.

Prochlorperazine (Compazine) is an anti-emetic normally used to treat nausea caused by radiation and chemotherapy, but sometimes prescribed for severe seasickness. Like Phenergan, it comes as a capsule, oral liquid, injection, or suppository. Some transoceanic cruising sailors carry suppositories in anticipation of encountering severe storms while underway.

Ondansetron (Zofran) is a relatively new and powerful anti-emetic intended for use by patients in chemotherapy. It doesn't cause drowsiness like Compazine and others (Hummel) (see note h). It is available as an orally dissolving tablet, which makes swallowing unnecessary. It is quite expensive, but few seasick individuals would balk at a $60 per pill price if it brought relief. As an aside, Dextroamphetamine (Dexedrine) is an amphetamine or stimulant that has been prescribed for seasickness, particularly in situations where alertness was essential. However, it is a controlled substance, has profound and sometimes dangerous side effects (Long 2003), and may be habit forming.

The armed forces reportedly supply specially concocted anti-motion sickness drugs to military pilots, astronauts, and some vessel crew members. One said to be highly effective is the "Navy cocktail," a blend of 25 mg each of Phenergan and ephedrine. Ephedrine is a powerful alkaloid stimulant included to counter drowsiness. A possibly safer blend has been suggested (Boniface 2003) that substitutes 60 mg of pseudoephedrine (Sudafed) for the ephedrine. Some fishermen use this combination, calling it the "Coast Guard cocktail."

Because drowsiness is a common side effect of many drugs, especially those based on antihistamines (in fact, similar antihistamines are packaged and sold as sleeping pills), some sources recommend taking a nonprescription caffeine-based stimulant, such as NoDoz, to counter the soporific effect.

Using drugs effectively

Research and field experience both show that there are two essential components to seasickness drug effectiveness. The first is to get a sufficient amount of the drug into the blood stream, and the other is to believe in it.

Just as individual physical size and personality varies from person to person, so does responsiveness to drugs. For some it may be necessary to use the manufacturer's recommended maximum dose, while others can cut it in half. When an already sick person takes a drug, it is difficult to determine how much of it has actually reached the system so it may be necessary to take more to reach the proper level. In most cases, if a drug isn't providing the expected relief, it is because an insufficient amount of it is getting into the system (see note i).

It is often said that OTC drugs such as dimenhydrinate and meclizine only prevent seasickness and do not stop it once it starts. This is not necessarily so. The problem is that once a person is sick, the pyloric valve (between stomach and small intestine) may close, preventing a swallowed pill from reaching the intestine where it can be absorbed into the blood stream. It probably will be expelled with the next bout of vomiting, and even if not, there is a delay of at least 2 hours while it works its way through and is absorbed so that it can do any good. That's why pills or capsules should be taken at least 2 hours prior to embarkation.

One charter boat captain has claimed great success with the following approach: at the first sign of queasiness, the sufferer takes chewable dimenhydrinate or meclizine tablets and chews but does not swallow. Instead the person holds the crushed tablets under the tongue or next to the cheek where it can be absorbed directly into the blood stream through the lining of the mouth (Maurice 2003). This captain reports that nearly everyone who takes these drugs by this *buccal* or *sublingual* route enjoys rapid relief. These drugs can irritate the mouth lining, and this method should be used only when really necessary. (Some researchers claim that this class of drugs cannot be absorbed through the lining of the mouth, but results of at least one clinical test show that dimenhydrinate can enter the bloodstream this way [Scavone et al. 1990]. Oddly, neither of the two primary manufacturers of chewable seasickness pills has tested this method of delivery.)

The other component of success is faith in the drug. Research shows that the placebo effect alone will alleviate symptoms in 40% of seasickness cases (Thornton 2002). When combined with the proven effects of selected drugs, this faith factor assures that most patients can be helped.

Conclusions

1. Mariners and passengers benefit from accurate information about sea-sickness causes, prevention, and treatments. Misinformation and unproven folk wisdom is in circulation.

2. There is no magic bullet. What works for some people does not work for others.

3. For many people, good preventative behavior can obviate the need for medication.

4. OTC motion sickness drugs tend to produce fewer and less intense side effects than prescription drugs, but also less of the desired result.

5. Among the OTC drugs, those based on meclizine are likely to cause less drowsiness than those based on dimenhydrinate.

6. The transdermal scopolamine "patch" appears to be the single most effective pharmaceutical solution for the broadest range of persons, especially in cases in which exposure to sea conditions is expected to continue for more than a few hours. Prospective users should be aware of potential side effects.

7. Mariners and passengers should assess their own susceptibility and plan ahead to prevent seasickness. Potential sufferers should "test drive" medications to check for side effects before going to sea.

8. Strategies for beating seasickness on day trips or short voyages may be different from those appropriate to longer voyages. For example, a strong dose of one of the common prescription anti-emetics would help an individual make it through a day trip, but it might interfere with the body's natural adaptation to motion, required for successful performance on an extended voyage.

9. Individuals who are drug-intolerant (and pregnant or nursing women) may prefer to try alternative remedies, such as powdered ginger and the acupressure or electronic acupressure devices.

10. For most people, the surest short-term prevention is to select one of the proven OTC drugs or the patch, take it early and correctly, and practice good preventative behavior.

11. Persons who have extreme susceptibility or anticipate extreme conditions should consider getting a prescription for one of the more powerful anti-emetics to be carried on board for use only in case of emergency.

12. Nonprescription chewable medications can be carried on board all vessels and offered to needy individuals who may wish to apply the sublingual/buccal absorption method.

13. Faith in a remedy is powerful medicine.

14. Seasick persons are sick and must be treated as such. They should be secured while vomiting and during rough weather to ensure that they don't fall overboard or become injured. Dehydration can be a serious consequence of seasickness in the short term (particularly in children), and nutrient depletion can be a consequence on longer voyages.

Suggestions for further research

Since most people embarking on voyages don't expect to get seasick, they don't take preventive measures. Future research should focus on measures to suppress seasickness after it has begun. Controlled studies on the effectiveness of taking chewables through the mouth lining, rather than swallowing, are warranted.

It would be useful to have further study of the effectiveness of combining measures, such as using acupressure bands in conjunction with drugs, and with combinations of drugs. Desperately seasick people often pile on the potential remedies, and it would be helpful to know if some advantage can be gained or if the potential risks are greater.

More controlled experiments in actual field conditions, such as aboard naval, research, fish processing, or work vessels, may produce more useful results than laboratory research, much of which is conducted in revolving chairs.

References

American Society of Health-System Pharmacists, Inc. (2001). *Drug Information*, Vol. 56, no. 22, pp. 2795-7, 2799-2805

Boh R and Anonymous (1999). Letters in *Latitude 38 Magazine*. Available at www.yachtsdelivered.com/seasick/stugeron.htm

Boniface K (2003). Seasickness: Coping with mal de mer. *The Quarterly*, 1st quarter. Available at www.vanhallhealth.com.

Cheung B (2003). Personal communication.

Cheung BSK, Money KE et al. (1990). Motion sickness susceptibility and aerobic fitness: A longitudinal study. *Aviation and Space Environmental Medicine* 61: 201-4.

Drouin M (2001). How to beat seasickness. *Pacific Fishing* May: 31-34.

Gahlinger P (1999). Motion sickness: How to help your patients avoid travel travail. *Postgraduate Medicine* 106(4).

Holtmann S, Clark AH et al. (2003). The anti-motion sickness mechanism of ginger: A comparative study with placebo and dimenhydrinate. Dept. of Otorhinolaryngology, Grosshadern Medical Center, Ludwig Maximilians University, Munich. Available at http://seasickness.co.uk/html/body_papers_40.html

Hummel C (2003). Personal communication.

Long PW (2003). Drug facts. Available at www.mentalhealth.com

Maurice M (2003). Seasick cure. Available at www.yachtsdelivered.com/seasick/sickcure.htm

Money KE, James R, et al. (1996). The autonomic nervous system and motion sickness. *Vestibular Autonomic Regulation*, Yates B and Miller A, eds. Boca Raton, FL: CRC Press, pp. 147-168.

Putcha L, Berens KL et al. (2002). Pharmaceutical use by US astronauts on space shuttle missions. Houston: Life Sciences Research Laboratories, NASA-Johnson Space Center.

Scavone JM, Luna BG et al. (1990). Diphenhydramine kinetics following intravenous, oral and sublingual dimenhydrinate administration. *Biopharmaceutics and Drug Disposition* 11(3):185-189. Abstract available at http://www.ncbi.nlm.hih.gov/entrez/query.fcgi?cmd=Retrieve&db=pubmed&dopt=Abstract&list_uids=2328304

Tassier P (2003). The artificial horizon glasses: A word from the inventor.

Thornton J (2002). Gut reaction. *Boating Magazine* April.

Notes

a. During his commercial fishing days, the author had a crew member who got desperately ill in the first storm of each season, and then was fine the rest of the year. That former crew member recently told the author that since he became a vessel skipper, he has had no further seasickness.

b. This subjective observation is based on the author's seven summers of taking tourists offshore on his small motor boat. When women get seasick they usually just deal with it; when men do they are more likely to articulate their condition, insist on curtailing the voyage, or even cancel altogether.

c. See, for example, Pitler and Ernst, "Efficacy of ginger for nausea and vomiting: A systematic review of randomized clinical trials," *British Journal of Anaesthesia* 84(3): 367-71. Found at www.jr2.ox.ac.uk/bandolier/booth/alternat/AT128.html. Also, J.J. Stewart., M.J. Wood, C.D. Wood and M.E. Mims, "Effects of ginger on motion sickness susceptibility and gastric function," Louisiana State University Medical Center, Shreveport; and A. Grontved, T. Brask, J. Kambskard and E. Hentze, "Ginger Root against seasickness: A controlled trial on the open sea," Dept. of Oto-Rhino-Laryngology, Svendborg Hospital, Denmark. Both abstracts are in *Medical Papers* and found at http://seasickness.co.uk/html/body_medical_papers.html.

d. The makers of Sea-Band claim "clinical trials" but do not provide traceable references. A company spokesman told the author that the company has difficulty demonstrating effectiveness on motion sickness because a great many variables come into play, unlike nausea related to surgery and other conditions that can be tested in the controlled environment of a hospital.

e. Lee Beall, a "volunteer seasick remedy tester," tried several devices for *Powerboat Reports*, an independent subscription-supported journal for boaters. Trials were conducted over a period of 6 days in a moving automobile to assure more consistent conditions than could be expected in a boat. In the October 2001 issue, in an article titled "Motion sickness remedies: Wrist bands worth a try," he is quoted as reporting that the two acupressure bands tested, Sea-Band and BioBand, both worked as advertised and prevented onset of motion sickness. In the same test he found dramamine very effective, while bonine and ginger tablets were ineffective.

f.   Beall, referenced above, reported in the same *Powerboat Reports* article that the Relief Band was highly effective in the automobile test, but was less so in an on-water test, and caused minor irritation to the wrist. John E. Phillips, writing in advertising-supported *Saltwater Sportsman,* reported successful results when he tested it on three charter boat anglers who were already experiencing some level of seasickness (found at www. saltwatersportsman.com/new_gear/sea_sick_cure.html). The author has observed aboard his own boat that about half of the people already feeling ill who tried Relief Band have enjoyed noticeable improvement. No-one aboard the author's boat has applied Relief Band as a prophylaxis.

g. For example, Robert Hoyt, MD, writing in the July 2000 issue of *South-winds–Local News for Southern Sailor,* in an article called "Seasickness ad nauseum," states, "I did find one relevant study that monitored about 1,400 patients on the open ocean, during which several common drugs were studied (cinnarizine, cyclizine, meclizine, dimenhydrinate, scopolamine and ginger). Their results showed that all the compounds studied had a similar effect in reducing the likelihood of seasickness from about 80% on no meds to 20%." He did not provide a citation for the study.

h. Dr. Hummel tested several medications on himself while on the author's boat and wrote a detailed letter describing his experience with effectiveness and side-effects.

i. This page describes how one skipper has adjusted doses offered to his passengers and crew members. Found at www.yachtsdelivered.com/seasick/dosage.htm.

DROWNINGS—
NO NEW CAUSES: COOPERATIVE APPROACH
NEEDED

Harvey A. Linton, MEd, BComm, Capt. 350T
Manager, General Industries Prevention Division
Workers' Compensation Board BC
PO Box 5350 Stn Terminal
Vancouver, British Columbia, Canada
E-mail: Hlinton@wcb.bc.ca

Introduction

People have gone to sea to harvest fish and other seafood since the beginning of time. Many did not return, perishing due to some misfortune related
to their venture. More often than not, the cause of death was drowning.
Men and women now go to sea to catch fish with modern vessels, advanced
navigation equipment, and first-class life-saving equipment. Regardless of
these improvements to vessels and equipment, drowning continues to be the
major cause of deaths. For example, in British Columbia, on average one
work-related death is recorded for every 26 workers' compensation claims in
the commercial fishing industry. Eighty-six percent of these were drownings.
The sad thing is, the same types of incidents occur over and over again. There
are no new causes. Change is needed, but change will not happen by itself.
All parties involved in, or with, the industry must focus and sustain their effort to reduce the number of drownings in the British Columbia commercial
fishing industry. This has begun to happen with formation of the Marine
Action Group in 2002.

This paper identifies and discusses—

- The factors contributing to drowning.
- The types of controls needed to prevent drownings.
- Who has direct and indirect responsibility in the industry.
- New cooperative initiatives being undertaken, including the emer-
  gence of the Marine Action Group.

The problem/issue

Between 1991 and 2002, 63 work-related deaths were recorded, 86% of which were drownings. This is despite the fact that the number of fishermen and women and vessels participating in the industry declined significantly. Drowning and other deaths occurred in all fisheries and gear types. Table 1 provides a breakdown by gear type.

Table 1: Drowning and other deaths by gear type

Gear	Drowning	Other	Total
Trap fishing	14	0	14
Dive fishing	9	1	10
Longlining	8	1	9
Gill-netting	7	3	10
Seining	6	1	7
Trawling	5	1	6
Trolling	4	1	5
Barnacle harvesting	2	0	2
TOTALS	55	8	63

Causes of drowning

The general causes of the various types of fatalities were determined by reviewing employer injury reports to the British Columbia Worker Compensation Board (WCB) for the period 1991–2001 and through contact with the WCB officers that investigated specific accidents. This information is summarized in Figure 1. It reaffirms that drowning is a very significant issue in the British Columbia fishing industry and that drowning results from two main causes: vessels capsizing or foundering, or crew members being dragged or knocked overboard and jumping or falling overboard.

A more in-depth look at contributing factors is necessary to prevent drownings. Reviews of accident investigations, incident reports, inquests, and other reports show that factors contributing to drowning generally fall into three main categories: structural, cultural, and insufficient information. These three categories are described below.

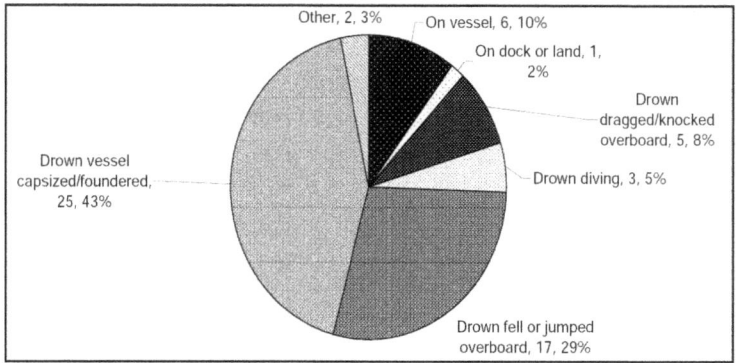

Figure 1: General causes of fishing deaths, 1991-2001

Structural factors

Structural factors include time pressures, lengths of seasons or openings, and quotas that must be filled within a specific timeframe. Vessel length and power restrictions may also be included. Such factors tend to increase economic or financial pressure such that vessel skippers and crew take chances they might not otherwise take. Frequently, situations arise from the need of a level of government to juggle and balance resource conservation and allowable catch with economic needs of the industry. A byproduct of increasingly reduced fishing time is that both skippers and crew members get less sea time and experience than they would have several decades ago.

Cultural factors

Cultural factors are often based on long-standing beliefs of people in the industry that they won't be victims of tragedies. They believe it won't happen to them because (in their words)—

- This boat has fished in these conditions for years with no problems.
- Working on deck with a PFD, harness, and lanyard would be more dangerous than not wearing the protective equipment.
- The feel of the boat provides all the stability information I need.
- I can travel safely after working round the clock because I have alarm systems and a good automatic pilot.
- This is the way that it has always been done in this industry.
- We have good boats and seamanship in this fishery, not like those guys in some other fisheries.
- Experienced fishermen and women are tough and have what it takes.

Insufficient information factors

Many incident reports reveal either a lack of information or a failure to consider, use, or obtain information needed to make sound decisions relative to the situation and conditions. Examples include—

Lack of information:
- Local knowledge of the impact of specific wind-opposing current conditions.
- Impact on stability from vessel modifications and changing gear type or fisheries.
- Design and special stability characteristics of the vessel under light and heavy loads.

Failure to use information:
- Proceeding because it should only take a few minutes to get through this hazardous situation.
- Continuing to catch and pack as much as possible because "There will never be another chance to catch this many fish."
- Proceeding through hazardous conditions because "We have to be at the dock unloading in 7 hours regardless of conditions."
- Believing that because a crew member has been in the industry for years, he or she is a competent watch-keeper.
- Believing that skippers and crew members who have been in the industry for years know what to do in the case of emergencies, such as fire on board or abandoning ship.

Cultural factors integrated with insufficient information can result in failure to analyze effectively probable and actual conditions and circumstances prior to making decisions that affect the safety of vessel and crew.

Controls to prevent drowning

Based on incident reports, four types of controls are needed to reduce the number of drownings. They are—

- Up-to-date, understandable, quickly read stability information for each vessel.
- Watch-keeping practices that include periodic monitoring by the skipper or a competent hand.

- Effective deck layout and work practices to prevent involuntary entry into the water and effective means of staying afloat if in the water.
- Periodic emergency drills for crew member overboard, fire, flooding, abandoning ship, and calling for help.

It is imperative that skippers and crew members be able to recognize the risks and minimize them quickly and effectively. To do this, they must understand the need for and fundamentals of each type of control identified above. The reasons why these controls are needed are as follows:

1. Fishing vessel stability has been a major concern for years in British Columbia as is evident from reviewing Transportation Safety Board reports. Between 1975 and 2001, 154 British Columbia fishing vessels capsized with 149 persons confirmed drowned or listed as missing. While these statistics do not distinguish between smaller open boats or skiffs and larger vessels, the fact remains that lack of sufficient stability has been a major factor leading to capsizing and drownings.

2. Inadequate watch-keeping has contributed to numerous vessels coming to grief through taking on water and foundering. In some cases, foundering resulted from serious leaks that developed in the hull. In other cases, flush deck fittings or hatch covers were not properly secured. Sometimes flooding occurred while in transit; at other times, it occurred after the vessel struck a submerged object or the shore. The fact is that in most cases, effective watch-keeping could prevent a flooding situation, provide time to control flooding, or provide time for a skipper and crew to prepare to abandon ship.

3. Staying out of the water sounds like an issue that wouldn't require discussion. However, WCB statistics show that a significant number of crew members drowned after they were dragged, knocked, or fell overboard. Typically, crew members do not wear personal flotation devices (PFDs) when working on deck. Even the light, self-inflating, low-profile types that don't impede work are seldom worn. Additionally, there are times when it is necessary to rig lifelines, and, in the case of trap fishing, wear a harness and retractable lanyard during some specific operation. All too often protective equipment is not rigged or worn because skippers and crew members assume or believe that it is too restrictive and they are likely to get caught in the gear. These cultural factors discourage searching for safer ways to work, contributing to inadequate protection and potentially increasing the risk of crew members going overboard.

4. Periodic emergency drills are a must. Generally, skippers and crew on larger vessels are much more attuned to conducting emergency drills than they were a decade ago. However, investigations have revealed a tendency on many vessels for both skippers and crew members to assume that individual skippers and crew members know how to respond effectively to all common types of emergencies based on their length of time in the industry.

Responsibilities for safety

Typically, when fatal accidents occur, there was a lack of understanding of and/or a failure to carry out responsibilities. Responsibility for safety in the fishing industry falls into two main categories, direct and indirect. These are described below.

Direct responsibilities include those of the vessel owner, skipper, and crew members. For example, the owner is responsible for maintaining the vessel in a mechanically sound and seaworthy condition. This would include documenting the impact on stability from vessel modifications and shifting to other gear types. The skipper is responsible for setting the operational standard for vessel seaworthiness and crew safety. Seaworthiness includes factors such as loading, tanking, and trim. Examples related to crew safety include watch-keeping, safe work practices, emergency preparedness, etc. Crew members have a responsibility to maintain themselves in fit condition and follow safe operating procedures, including wearing protective equipment and applying safe work practices.

Government or government agencies have *secondary or indirect responsibilities* for safety. A number of jurisdictions are involved. Transport Canada has marine responsibilities under the Canada Shipping Act. In practice their focal areas include vessel structure and the maintenance and operation of vessels according to international and specific conventions. The WCB sets and enforces minimum safe work standards and may investigate accidents. The Department of Fisheries and Oceans licenses vessels and crew, and administers the opening and closing of fisheries and areas relative to sustainability of stocks. The Canadian Coast Guard has responsibility for search and rescue. The Transportation Safety Board investigates marine occurrences.

Many of the factors discussed earlier in this paper, to varying degrees, have singly or in combination with responsibility issues contributed to drownings in each gear and fishery type throughout the years. To prevent these drownings, both those with direct and indirect responsibilities for safety in the industry must take progressive, focused action. The fundamental steps in the process are to—

+ Identify priority risk areas, specific causes, and contributing factors.
+ Build awareness of specific risks, causes, and contributing factors that frequently result in drowning.
+ Build on awareness to generate timely recognition of when and how to apply preventive measures.
+ Ensure compliance with applicable crew and vessel safety regulations.

What is being done?

Many individuals and organizations with safety responsibilities within the industry have recognized causes and contributing factors that increase the risk of drowning and have taken steps to reduce the risk. For example, some vessel and fleet owners put their skippers and crews through Marine Emergency Duty (MED) training and hold preseason safety meetings. Some marine insurance companies have raised their safety standards for insured vessels and hold periodic safety forums. The Transportation Safety Board investigates some serious fishing vessel accidents and publishes detailed reports of these accidents. Transport Canada inspects vessels over 15 tons every 4 years. The Department of Fisheries and Oceans has begun to factor weather and other conditions into openings and length of time to fill quotas. The Coast Guard engages in search and rescue may provide inspection and education services. The WCB inspects conditions covered under its regulations, investigates accidents, and provides consultation, education and various publications.

What else needs to be done?

Given the geographically dispersed, highly mobile, and multicultural nature of the British Columbia fishing industry, a catalyst was and is needed to encourage and promote a greater degree of collaboration on issues, initiatives, and information sharing.

Clearly, numerous individual initiatives are aimed at both crew and vessel safety. However, to achieve a sustainable reduction in drownings, a broader sharing of information and services is needed through sustained collaborative efforts that focus on causes and contributing factors.

As described above, a more cooperative spirit between industry organizations, individuals, and government has begun to emerge. There is now some room for optimism that partnering and sharing information, services, and publications will generate sufficient synergy for a semi-formal fishing industry safety program to continue to evolve.

The emergence of the Marine Action Group

The Marine Action Group (MAG) emerged through meetings between the WCB and the four federal government marine agencies in 2002. MAG was formed to share information about priority issues, initiatives, and educational materials. By late 2002, it became apparent that industry organizations should also be invited to participate. The BC Seafood Alliance was the first industry organization to become involved. Early in 2003, the Council of Professional Fish Harvesters began to participate. The United Fishermen and Allied Workers Union and the Pacific Coast Fishermen's Mutual Marine Insurance Company, the largest insurer of fishing vessels, have received invitations and are expected to participate.

MAG's aim is to provide an opportunity for organizations to broaden distribution of their educational safety materials to all segments of the industry. Some initiatives will remain the domain of the individual organization, and others may be developed through joint agency/industry organization working groups. The common purpose is to create a broad awareness of priority safety issues, such as drowning, and necessary behavior change within fishing communities and on board vessels. The process includes generating—

- Awareness of individual and collective responsibilities for preventive action and activities.
- Awareness through promotions, posters, presentations, industry leader testimonials, discussion groups, and enforcement.
- Increased use of naval architects when vessels are modified or rigged for a different gear type.
- Increased numbers of vessel masters ensuring their crews complete MED training.

+ Application of effective means to recover a person overboard on every vessel.
+ Improved deck layout and application of other means on board to reduce incidents of crew members being dragged, knocked, or falling overboard.

Activities, services and publications include, but are not limited, to—

+ Fish Harvesting Alert (fatality poster).
+ Hazard Alerts.
+ Fishing vessel stability poster (*new*–illustrating the impact of modifications and changing gear types on vessel stability).
+ Funding an industry safety coordinator position.
+ Monthly articles in fishing journals.
+ Funding (partial) the May Day fishing safety project for exhibit in coastal museums.
+ "Gearing Up for Safety" manual and other manuals.
+ Presentations, inspections, consultations, and enforcement by the WCB and other organizations.
+ Investigation of the *Cap Rouge II* capsizing by the Transport Safety Board and WCB.
+ Use of laser technology by a naval architect to capture the lines and calculate stability of the *Cap Rouge II* under various assumptions and conditions.

Agencies and organizations participate in MAG because they believe collaboration and partnerships provide an opportunity to achieve greater results. While the WCB and others would like to see no drownings within 5 years or less, a 75% sustained reduction within the same period would be considered a successful beginning. The WCB would also like to see the death-to-compensation claims ratio improve from its current position of the worst ratio among all industries in British Columbia. Perhaps the greatest challenge for MAG and participating agencies and organizations will be to maintain focus on generating results and resist becoming a forum for meeting goers.

Evaluation

As previously mentioned, formation of the MAG and related activities are emerging as a semi-formal program that focuses preventive activity on reducing drownings and other serious incidents in the fishing industry. At this time, a formal evaluation process has not been discussed. However, it is anticipated that an evaluation process will be developed in which levels of preventive activity, behavior change, and results are measured. Activities such as promotions, presentations, educational forums, consultations, and publications will be discussed in MAG meetings along with recipient numbers. MAG working groups will also develop and deliver presentations and materials.

Activity effectiveness will typically be evaluated by the organization providing the service or publications and communicated to other MAG members for discussion where pertinent. Routine inspections and accident investigations by participating organizations will indicate whether or not behavior change is occurring in the industry. Surveys on awareness of causes and application of preventive practices on board vessels may also be considered. Transportation Safety Board statistics, WCB accident statistics, and WCB compensation claims statistics will be reviewed periodically to determine progress toward reducing the average number of drownings and the need to focus on new fishery groups, causes, or contributing factors.

Summary and conclusions

There is no reason for the unacceptable number of drownings in the British Columbia fishing industry to continue. The causes and factors that contribute to the majority of drownings are known. Many organizations and individuals in the industry have risen to the challenge and taken the initiative to reduce or eliminate drowning. A much higher level of cooperation and collaboration now between the WCB, government marine agencies and industry organizations to increase awareness of causes, contributing factors and means of prevention. The evolution of a semi-formal industry safety program to engage fishermen and women and fishing communities is gaining momentum. Through collective, progressive action that continues to engage all stakeholders, the goal of zero drownings will be realized. Return to the status quo is simply not an option.